Olympi

MW00761957

7.31% $\int 6t^9 \, dt$ $(8x+1)^2$ 7.31% $\int 6t^9 \, dt$ $(8x+1$

$\int 6t^9 \, dt$ 7.31% $\int 6t^9$

7.31%

$\int 6t^9 \, dt$

$(8x+1)^2$ 7.31% $\int 6t^9$

7.31% $\int 6t^9 \, dt$ dt $(8x+1$

$(8x+1)^2$ $)^2$ 7.31

$(8x+1)^2$ $\int 6t^9$

7.31% x+1

$6t^9 \, dt$ .31

$8x+$ ^{9}

.3

$6t$

$8x+1)^2$

How You Can Overcome Your Math Fears and Become a Rocket Scientist

MATHAPHOBIA

Praises for *Mathaphobia*

"Olympia's book *Mathaphobia®: How You Can Overcome Your Math Fears and Become a Rocket Scientist* is inspiring, and is needed in American schools. Adults, parents, college students, and pre-college students will benefit from reading this book. I support Olympia in her passion to end *Mathaphobia®* in America, and I ask that you help Olympia LePoint in this S.T.E.M. education journey."

-- Robert Curbeam, Retired U.S. Astronaut

"Olympia LePoint is a gifted academician and practitioner with an innovative approach to improving mathematics literacy, and dispelling the crippling fears that are often associated with learning the subject. In *Mathaphobia®: How You Can Overcome Your Math Fears and Become a Rocket Scientist*, Olympia offers powerful tools and effective strategies to help readers identify and overcome the mental roadblocks to excelling, not only in mathematics, but also in life. She combines important results from human psychology with solid, practical insights from her professional and life experiences to create this masterful and highly accessible volume. As a fellow educator, I have used many of her principles to empower people of all ages, backgrounds, and abilities to overcome their math fears. Teachers, students, and parents will be able to quickly apply Olympia's simple, three-step success plan to thinking more clearly, critically, and confidently. *Mathaphobia®* also tells a beautiful story of Olympia's journey from humble beginnings in inner-city Los Angeles, to the heights of excellence at the Boeing Company. Olympia has overcome great odds and personal challenges to excel at her profession. Her story demonstrates profoundly that with faith, discipline, and courage, there is no fear that cannot be conquered and no trial that cannot be overcome. *Mathaphobia®* is a very significant work that will truly inspire men, women, and people of all ages to become rocket scientists and other exciting things. It is a great honor and privilege to be able to recommend it to you."

-- Reverend Wilfred Graves Jr., Ph.D.
Statistician, Former Aerospace Engineer Author,
In Pursuit of Wholeness: Experiencing God's Salvation for the Total Person

Praises for *Mathaphobia* Continued

"I have known Olympia LePoint for about 15 years. After first meeting with Olympia LePoint as a young engineer and mathematician in 1998, I was highly impressed with her knowledge and enthusiasm on mathematics and engineering. While she worked for our leading American rocket propulsion company, she constantly learned and applied mathematics and statistics techniques to help design rocket engines and launch rocket engines safely. Now, I am really glad to see Olympia brings her real world knowledge and experiences to students, so they can succeed in Math, Science, Technology, Engineering and Math (STEM) fields. I highly recommend this book for everyone who wants to overcome their math fears. The book shows easy ways to study math, and how to connect with teachers. This book brings me great memories of how Olympia presented highly complex math and engineering problems to NASA and Industry customers. I am confident that this book will help students succeed in the same way."

-- Dr. Zhaofeng Huang,
Technical Fellow
A leading rocket engine company

"Olympia LePoint's description of her psychological phobias of Math literacy, calculations, and analytical problems are easy for all of us to connect with. Her storytelling approach is a synthesized blend of Brain 101 and Math 101, which describe possible self-sabotaging thought patterns. These hindering patterns are responsible for Innumeracy (the shutdown of the brain's frontal lobes, and the block of innovative thought). She brilliantly correlates personality types with the inappropriate responses, actions, and beliefs as she describes four types of mathaphobic people. I was able to personally go from *Samuel* the Struggler to Ivan the Innovator, showing that overcoming '*Mathaphobia*®' can be as simple as embracing a new set of tactics and studying differently."

--Dr. Dale Deardorff
Director of Innovation & Strategic Thinking
Rocky Peak Leadership Center

Praises for *Mathaphobia* Continued

"Olympia has been able to emerge from humble beginnings with a drive, inquisitive nature, and a thirst to learn that enabled her to succeed at reliability analysis on advanced rocket propulsion systems. She has leveraged this energy and desire to help others to create the book *Mathaphobia®: How You Can Overcome Your Math Fears and Become a Rocket Scientist*. As I worked with her on improving the safety and reliability of our future rocket engines, she taught me to continually drive towards my dreams and better understand the ways to lead effectively. Olympia LePoint has captured thrilling rocket engine examples in the book *Mathaphobia®*. Packed with insight and wisdom, *Mathaphobia®* will help countless future Rocket Scientists, and adults, to conquer their fears of mathematics."

-- Stephen Hobart,
Rocket Engine & Advanced Propulsion Development Manager
A leading rocket engine company

"In 2007, Olympia LePoint was hired to teach math at Pierce College. After only a few weeks of teaching a Trigonometry class, students began to repeatedly request Olympia to teach their future math classes on campus. Olympia LePoint has taught math for over five years at our campus. With innovative math study skills approaches, Olympia offered to teach Math Jam Workshops on the campus. These classes were geared for students who needed a math refresher course before their Fall classes began. She shared her *Mathaphobia®* removal process for students. We received outstanding reviews from each participating student. The students were excited about the success tips they acquired from Olympia. She held these supplemental workshops for a number of years on the campus. After working to capture her recommendations on paper, now Olympia has incorporated these strategies into her book *Mathaphobia®: How You Can Overcome Your Math Fears and Become a Rocket Scientist*. Olympia is a supportive, entertaining and informative instructor who has the ability to get inside students' minds to translate technical concepts in clear ways. Able to cleverly motivate each student to pursue success, Olympia is a well-liked educator. As a result, I am positive that Olympia's book *Mathaphobia®* will help many students to succeed in math."

-- Robert Martinez,
Math Instructor and Math Department Chair,
Pierce Community College of Los Angeles Community College District

MATHAPHOBIA

**How You Can Overcome Your Math Fears
and Become a Rocket Scientist**

Olympia LePoint, M.S.

OL Consulting Corporation & Publishing
Los Angeles, California

Mathaphobia®: How You Can Overcome Your Math Fears and Become a Rocket Scientist
Copyright© 2013 by Olympia LePoint, M.S.
Mathaphobia® Registered Trademark - O.L. Consulting Corporation

All rights reserved. Printed in the United States of America

All rights reserved. No part of this book may be reproduced by any mechanical, photographic, or electronic process, or in the form of a phonographic recording; nor may it be stored in a retrieval system, transmitted, or otherwise be copied for public or private use without written permission from the publisher. For information regarding permission or additional copies, contact the publisher.

Published by: O.L. Consulting Corporation & Publishing

Los Angeles, California 91325
818.775.9871

Address all inquires to:
Inquiries@OLConsultingCorp.com

Includes bibliographical references.

LCCN: 2012924113
ISBN: 978-0-9885376-3-7 (Paperback)
ISBN: 978-0-9885376-8-2 (eBook)

Editor: Penny Scott
Cover Design and Page Layout: Roman Warren
Cover Photo: Mark Robert Halper
About the Author Photo: Valeria Papp
Back Cover Photo: Todd Tyler
Mathaphobia Character Illustrator: Rayfield Angrum

I dedicate this book to my mother Pamela LePoint. I thank you for being here on Earth with me after July 8, 2004.

I am honored to be half of you.

Acknowledgments

I thank my mother, Pamela LePoint, who stands by my side during my trials, and success. I am honored to be her daughter. I'd also like to thank my family members.

True friends are the ones by your side when the limelight is off. I'd like to thank the following people: Cristine Carrier and Bodo Schmidt, Ed and Laurie Carmack, Adam Nunn, Brooke and Tom Turpin, Jeanine Powell, Diana and William Zelnis, LaToya Brown, Aileen Duenez, Tyrone Fox, Marie and Laurie Lundin, and Justin Phillips. I've been fortunate to have met a number of great mentors, thank you especially to the following: Bonnie Collins, Paul Richey, Ben and Cindy Tenn, Nick Johns, Phil Monroe, and Shellie Hadvina. I am privileged to have great pastors in my life, as well as the following university professors: Dr. William Watkins, Vice President of CSUN, along with Dr. Mark Schilling, Dr. Frankline Augustin, Dr. Michael Neubauer, Dr. Werner Horn, and the late Mrs. Jane Pinkerton.

I would also like to thank my Rotarian friends: Don Robinson, Maureen Tepedino, Gerry and Kathy Turner, LJ Rivera III, Jim Hoyt, Carol and Bob Donahue, Denecia Jones, Beatriz Phipps, and Penny von Kalinowski. And I thank my *Mathaphobia*® Educator Staff: Michael Carlson, Roger Richardson, Shea and Courtney Smith, Melanie M. Aunchman and Silva Feroian.

There are several co-workers who've helped me, including; Dr. Zhaofeng Huang, Dr. Dale Deardorff, Steve Hobart, Sandra J. Washington, Dennis Early, Bob Martinez, Carl Main, as well as my special astronaut friend, Robert Curbeam (retired U.S. Astronaut). I also thank Ellen Faitelson and Dirk Zirbel's family, Ana Bonilla, and Chryssi and John Trachsel, Kael Beverly, and Terresa Burgess.

Each of them re-encourages me to overcome situations when Life throws every unimaginable challenge. I am thankful to have built relationships with these honorable men and women.

And, most of all, I am grateful for one moment in time related to this book. One night, I went to sleep asking the universe for a way to complete this masterpiece: given my situation. When I awoke, I received clear direction to call my Freshman English Professor, Johnie Scott. Perplexed, I followed this instruction. After calling him, I was floored to discover that his daughter was a seasoned book and magazine editor, who would later be the editor of this book. As a result, I thank God for introducing me to Angela "Penny" Scott, who helped bring my stories to life!

Table of Contents

Introduction: xx

Chapter 1: I'm a Rocket Scientist?.. 2

Chapter 2: Do We Really Need Math?...................................... 19

Chapter 3: Do You "Math – Sabotage" Yourself?..................... 35

Chapter 4: Open Your Heart = Open Your Brain 59

Chapter 5: Capitalize on Your Own Thinking Power................. 75

Chapter 6: Turn *Quincy* the Quitter into David the Determined............ 99

Chapter 7: Turn *Donna* the Over-Doer into Sarah the Strategist 117

Chapter 8: Turn *Samuel* the Struggler into Ivan the Innovator 135

Chapter 9: Turn *Crystal* the Criticizer into Ellen the Explorer 161

Chapter 10: Lead through Change: Tips for Leaders,
Parents and Educators.. 179

Chapter 11: Get Up. Rise Up. Move forward. Go............................. 195

Foreword

When Olympia LePoint mentioned to me that she was writing a book called *Mathaphobia®: How You Can Overcome Your Math Fears and Become a Rocket Scientist,* I was hopeful yet wondering how a book could be written to truly help those in need.

My own background has been as an executive in healthcare operations, where I have overseen multi-million dollar budgets and thousands of employees. I have seen firsthand, where staff has been uncomfortable dealing with Profit & Loss statements, basic percentages, and analyzing budgets. When I've inquired, many employees stated they had a fear of "numbers," which oftentimes stemmed from early negative experiences with math.

Even during my doctorate studies in Psychology, many of the graduate students dreaded taking the advanced statistics and quantitative methods classes. Some of my colleagues never finished their Ph.D. due to the extensive research and analysis required for the thesis. While my own career and education have been successful, I too struggled with math. I had to use all of my resources and faculties to muster through my college math courses. If anyone can appreciate mathphobia® and trying to overcome it, then I would be your resource. I can certainly recommend Olympia LePoint's *Mathaphobia®: How You Can Overcome Your Math Fears and Become a Rocket Scientist* book as a great first step in overcoming math fear and mastering the subject.

I was first introduced to Olympia by my wife, as a result of my two children, Jessica and Zachary needing help with their math skills. We had heard about a scientist turned motivational math instructor, and decided that our own math skills were stale. We needed a true professional to assist our children. Olympia began helping my children over five years ago. My daughter was the first one to start with Olympia, and gradually grew more and more confident in her abilities to solve math problems. She eventually took Chemistry and other advanced math courses, and has since moved on to attend college. Jessica is now in her second year of college, well on her way to becoming a college graduate! My son has made great strides in his understanding and comfort level of math, and he is in his junior year of high school now. Zachary is deciding upon colleges, with a newfound level of confidence as a result of Olympia's step-by-step method outlined in her *Mathaphobia*® book.

Those of us that have taken math have both good and bad experiences: some as a result of the material, and other negative experiences due to a poor instructor. Olympia LePoint's *Mathaphobia*® book addresses the psychological and physiological impacts on those who suffer from mathaphobia. Olympia takes the reader on a tour of how to first admit to one's fears, and then how to take the journey to success. The book allows the reader to take specific steps to help eliminate fear, and then to ultimately make math fun. My own children began to use the book's principles, and I could actually hear them laughing with Olympia as she helped them progress from fear to success. Who would ever think that kids would laugh during math? It was simply amazing to see the transformation.

This is a remarkable book, a motivational self-help and instructional guide all rolled into one. Olympia LePoint has taken her skills as a mathematician, college educator to adults, teens, and college students and assimilated them into a book that will help thousands of people overcome their fear of math.

For those of us who have children, and also connected to the business and university worlds, I have no doubt that Olympia LePoint's book *Mathaphobia®: How You Can Overcome Your Math Fears and Become a Rocket Scientist* will help in all areas.

Dirk Zirbel, Ph.D.
Business Executive, and Proud Father of
the Students Olympia has Mentored

Introduction
The Birth of *Mathaphobia*®

If you can remember your previous math classes, you will have one of two emotions: happiness or pure terror. Unfortunately, pure terror is the response that most Americans have toward their own math experience. Math fear has become a national phenomenon that is causing record numbers of hard-working individuals to turn away from their dream careers, dread school, and feel defeated. Countless require some form of intervention: They have either forgotten how to perform basic math; are afraid of failing math tests; or are in desperate need of education strategies. I personally know this math fear phenomena, because this fear also crippled me. This fear nearly discouraged me from pursuing my dreams! I know I wouldn't have become a university graduate and an award-winning rocket scientist if I did not remove this fear from my life.

Now, math fear has a specific name.

When you can name and call out your math fear, you have the opportunity to overcome the emotion that seeks to paralyze you. When you are armed with effective knowledge, you win the best weapon to fight off math fear for good. My plan is to bring you enlightening information, so you can fight off the enemy that seeks to steal your future success. I am with you in this battle. And, I will show the strategies to fight back, because we are in this journey together.

I created this book, Mathaphobia®: How You Can Overcome Your Math Fears to Become a Rocket Scientist, after I founded the educational program, *Olympia's End Mathaphobia*™ Now. Before my experience as a rocket scientist, motivational speaker, and college educator, I had to first overcome my own fear of math. As a teenager, I would panic in math classes, especially during math tests! My body would cringe, and I'd sweat profusely. I had trouble solving for "x." My brain would flip somersaults like an Olympic gymnast. I repeatedly failed math.

However, while in college, I learned to conquer that phobia. I excelled, transforming into a professional, award-winning rocket scientist and educator. All of this occurred, once I ended my own self-sabotaging thought patterns toward math. After nearly 10 years of rocket engineering (from 1998 to 2007), I had an epiphany while sitting in the Mission Control support briefing room at Rocketdyne in Canoga Park, Calif. One day, I scanned the room. While gazing at a bunch of graying men in dull business suits and pocket protectors, it hit me. BAM!! This epiphany hit me like a ton of bricks! It was then I realized I would be one of the few remaining engineers. Those "suits" were scheduled to retire in the next five years. Not to mention, I was one of only a handful of women scientists.

It also dawned on me there would be no Americans to neither guard our national safety, nor participate in Top Secret space exploration. This shortage existed primarily because Americans dreaded math. Nonetheless, it wasn't until a year after my epiphany, when I decided to help secure our nation's technical future by training the next generation of U.S. engineers and scientists. However, once I got my feet wet, I realized this task was much more complicated than it first seemingly appeared. As a Los Angeles Community College Math Educator, it became apparent that many are math illiterate due to a type of mental virus. Mothers, daughters, fathers and sons all struggle with severe math fears. Adults are shying away from math in record numbers! Unfortunately, these adults are spreading this anxiety to their children. The culprit is what I've coined Mathaphobia®.

It is a contagious, mental virus and severe, math fear that blocks individuals from math literacy, basic math calculations, and analytical problem solving. Consequently, we are witnessing record numbers turn away from technical fields, which are increasing our U.S. financial decline. Surprisingly, mathaphobia® stems from three apparently unrelated sources: parental math fear, math–illiterate schoolteachers, and failing school systems. Thus, I created *Olympia's End Mathaphobia*™ *Now*, a math phobia eradication service for college, pre-college and returning college students. Through the system, I define, measure, and eliminate mathaphobia®. In this way, math-literate, innovative people can build our nation's future.

Olympia's End Mathaphobia™ *Now* has been widely accepted on California college campuses. Program workshops have been held at The California Science Center (Los Angeles); California State University Northridge; the University of California Los Angeles Extension Program; Pierce College of The Los Angeles Community College District; and at Santa Monica Community College. Here, they created podcasts from my *Math Made Real* presentation on Nov. 12, 2009. Contents of my math program are also enclosed within this book, Mathaphobia®. Nationally, I am interested in sharing this information through speaking, writing and consultation services. Together, I know that we can end Mathaphobia® for good. Now our mission begins. Are you ready?

I'm a Rocket Scientist?

CHAPTER 1

I'M A ROCKET SCIENTIST?

Do you think it takes a genius to be a rocket scientist? The answer is no. I will share a secret with you. There is a self-empowering process that enables anyone to become brilliant in math. Trust me. Becoming a rocket scientist has almost nothing to do with math. I would have never imagined myself as a rocket scientist. My background was the most farthest away from prestigious. In fact, I would argue that my development was extremely dysfunctional! Nevertheless, I was given the talent to imagine. I call it faith. If I could become a rocket scientist, given such early dysfunction, I am convinced that anyone can use her skills and talents to achieve her heart's desire. Of course, boys can succeed, too!

My motto is: When given lemons, make lemon meringue pie. Here's a better illustration about my "great mathematician preparation." Picture this scenario. My childhood was the 100 percent – complete opposite from "rocket science" performance. I was never on the Ivy League-educational track. And yet I became a rocket scientist! I was raised in South Los Angeles, Calif., with three sisters in a financially-and emotionally-depressed, inner-city neighborhood. Sadly, many of today's early learners are conditioned to have dysfunctional thoughts. Throughout the day, little ones are exposed to inappropriate behaviors exhibited through; movies, TV, discouraging teachers, poor schools, unsafe neighborhoods, and even unsafe environments at home. I'm no exception to these unfortunate circumstances.

My mother was a single parent, who struggled to pay the bills. Between the factors of poverty, gang violence, and educational ignorance, I was bombarded with anxiety from every angle. As a kid, I was safe, while caged between the four walls at home. Once I exited my house, I was surrounded by gangs and gang violence!

I was raised on 55th Street, between Hoover and Vermont, which was highly gang-infested during the late 1980s. I remember an "unspoken" curfew on Fridays, where everyone had to be inside by noon. If you were caught outside; male or female; despite nationality; wearing the wrong colors or shoes; you would be shot dead! Fridays were designated for weekly gang initiations. We even slept a certain way in the bed at night to ensure that bullets would miss our heads, if they penetrated the walls.

I also worried whether we had enough food for us as a family to survive. My mother always said, "The only way to get out of this poverty is through an education." So, I didn't want to let myself or my mother down by not earning a bachelor's degree. My number one goal was to make it out the 'hood'.

> *"Once I exited my house, I was surrounded by gangs and gang violence!"*

While growing up, my family had a lemon tree in the backyard of our home. A plethora of sour fruit fell from this tree. Even though sour lemons filled our backyard, I managed to make lemonade. Thanks to timing and the right directions, I graduated to making lemon meringue pies. All the same, my childhood was equivalent to a bucket of sour experiences. Fortunately, within time, I learned to make a great dessert from the lemons I received. Would you believe I used math to make the pies?

At 8-years-old, I first learned math, while doubling fractions for making lemon meringue pies and oatmeal cookie recipes. I am convinced that baking allowed me to master fractions and pattern relationships. As I covered fractions in school, I began to enjoy the ability to solve math puzzles. To my amazement, I excelled in math concepts, and was placed in a gifted class. I took pride, for I was one of the few 8-year-old girls, who knew how to quickly solve math problems. For me, math was an escape from life's chaos. I often felt bad going to my mom for help, after realizing that she, too, needed help raising us. With one older sister and two under me, math became my "imaginary brother." Math was a "know it all." We would argue all the time. Of course, he would prove how he was right. Math offered a definite structure when I desperately sought order and boundaries.

But, no amount of protection could save me from what happened inside of Mrs. Breland's fifth-grade classroom. She sat Glen, a troubled, 10-year-old boy from my neighborhood, next to me. I asked Mrs. Breland to move our seats, because I had a bad feeling about sitting by Glen. He was a tall, slender, caramel-colored African-American male known for wearing light blue L.A. Dodger's™ ball caps, sagging, creased dark blue jeans, and white T-shirts. Unbeknownst to me, Glen had just been recruited into one of the local street gangs. Recently, he started wearing this filed-down ring on his right, wedding ring finger. The spiked edges were sharp like a knife. Daily, Glen would taunt and tease me. I defended myself by playing psychological mind games on him. One day, Glen had heard enough of my smart mouth.

Glen snatched the class assignment out of my hands, ripping it up with a burst of pent up emotions. I heckled back, "Ooh, big man! All I have to do is get another piece of paper. And, write my answers down again. They're all in my brain!" That was enough for Glen. He was furious! Before I knew anything else, Glen socked me under my left eye. While punching my face, Glen used the ring on that right fist to slash open my left eye socket!

Immediately, I wanted to stand up and fight back!!

But, everything went black!! I was temporarily blinded in both eyes. I felt a cold, wet liquid, running down my face. I heard screams from my classmates because blood was squirting on them, tables and chairs. Needless to say, I was rushed to the hospital, where I received five layers of stitches. I was told that if the cut on my face was any higher, I would've lost my eye.

After that incident, my mother took me out of school. I was home-schooled for the remainder of the year. Several weeks later, I received a letter notifying me that I had been accepted into a gifted, academic magnet school for seventh grade. However, this posed a dilemma: I would not only be "skipping" a grade, but also attending school on the other side of town. Nonetheless, my mother ensured me this decision was for my safety, and to better my education. To say the least, my transition into this new gifted middle school was considerably overwhelming.

First off, I had always been raised in a predominately African-American and Latino-American environment. Alas, I learned that I wasn't speaking proper English. It was a complete culture shock for me! Here I am, going to school with holes in my shoes, and these kids from West L.A. and Beverly Hills, naturally assume that I would have the basic school supplies. I remember constantly borrowing paper from my classmates. That's just terrible and embarrassing. No child should have to be looked down upon for needing such items!

To make matters worse, I found out I was actually four years behind the educational curve. That was a BIG wake-up call for this scrawny, short kid from South L.A.! I was two years younger than my peers, and placed in a seventh-grade gifted program, where they were working on ninth-grade curriculum. It was a real rough period. Needless to say, this transitional time was more than turbulent. My early years became even more challenging, and I received my first *FAIL* in Algebra I. At 10, my accumulated environmental fears began to re-channel themselves into one big BEAST with a particular name. This severe math worry crippled me, and the phobia was my crutch for years. For the naysayer, this horrific experience debunks the assumption that I always excelled in mathematics. Surprisingly, there is an ironic part. I still loved math despite these circumstances. Math was my brother, and I knew we were destined to survive together!

With a little fire under my belly, I decided to stretch my wings and pursue performing arts. I was accepted into the Musical Theater Academy at Alexander Hamilton High School, which is nationally recognized for its top-notch music programs. Although math was my buddy, I still couldn't pull an "A" in any math class. Like many students, I couldn't comprehend what I was being asked to do on the math tests. Chemistry was another subject, where I failed to spend enough time figuring out its applications and equations. Performing in numerous school productions took my mind off of the trouble I was facing at home and in my math classes.

When it came time for me to consider attending college, I was ill-prepared. At the time, my mother lacked a higher education and couldn't offer any guidance. Like any great actress, I learned to mimic those around me. So, I applied to colleges and took the college entrance exams. I had average SAT scores, but my cumulative GPA was competitive enough to allow me acceptance into a number of colleges, including California State University Northridge (CSUN). I had the inner fire to get myself out of the 'hood,' and that's made all the difference. While attending CSUN, I received the ANSWER. To pay for school, I became a math tutor under the advisement of my late math mentor, Mrs. Jane Pinkerton. By working with hundreds of students, I witnessed that I, Olympia LePoint, was one of millions with the same math phobia.

Unfortunately, before working at Rocketdyne with some of the brightest space engineers, I suffered from this anti-math behavior as a teen and young adult. Fear had become my best friend. Then one day, this "terror" finally left, because I was given a special thinking gift. As soon as I learned the method to achieve in math, and in other university subjects, a number of my peers and employers were puzzled. They asked, "How did you make it out of the *'hood'*?" Deep down, I was extremely offended. For years, I hated this question. Mainly, because I erroneously thought I never owed anyone an explanation. However, I owed it to myself. My internal annoyance existed because I, in no way, truly investigated the reason. After years of pondering, I now know the answer. Are you ready? I will share the answer with you.

The answer is *MATH **and*** its creative problem-solving principles.

Can you remember your experiences in your past math classes? Most people do. Whenever I mention that I am a math professor, individuals usually cringe and tell me about their trauma while learning math. They either had difficult teachers or unresolved issues surrounding math. After teaching math to tens of thousands of students, I've witnessed far too many Americans suffering from math illiteracy, also widely known as ***innumeracy***. Through my research and experiences, I discovered that innumeracy is caused by one root culprit: *mathaphobia*®.

This fear is a contagious, psychological virus that blocks the brain from math literacy, basic math calculations, and analytical problem solving. Like other phobias, *mathaphobia*® shuts down the frontal brain lobes – the same lobes that are responsible for creative thinking.

The frontal lobes are needed to create math solutions. In turn, mathaphobia® activates the *flight–or-fight* survival response. If an antidote is not present, the mathaphobia® virus viciously spreads from person-to-person like a contagious plague. I saw my own fears through each person whom I helped.

> *"Like other phobias, mathaphobia® shuts down the frontal brain lobes –the same lobes that are responsible for creative thinking."*

I was forced to identify and eliminate my own self-sabotaging thought patterns through tutoring students with the same fear. Through time, I chose to relearn the math as I sat with each person. We learned the foundations together. I began to understand the intricate details of the universal math language. I saw that there is a sequential process to learn math and solve problems. Miraculously, my thoughts toward math changed, and my newly-gained confidence helped to resolve my real-life problems. With my newly-acquired critical thinking skills, (which I will show you in Chapter 4), I solved my real-life problems, and beat all odds against my favor.

In this book, I offer a fresh way to view math, expose the four types of mathaphobia®, explain the psychological thinking behind the mathaphobia® phenomena, and offer ways for parents, educators and students to become successful in math education. The three-step success plan removes your mathaphobia®, reprograms your brain, and makes you comfortable when solving math problems. As a result, your problem-solving abilities will transform into daily critical-thinking skills and self-empowerment.

Me, a Rocket Scientist?

My college students constantly ask, "Did you always want to be a rocket scientist?"

The answer is, "Yes." And after I think about my journey, I say, "No." I was 8 when I decided to become a rocket scientist. But, my dream was buried under years of subsequent disappointment. One morning, I found myself wide awake at 3 a.m. As a child, who never had an issue falling asleep at the drop of a hat, I was surprised when I could not return to sleep. To entertain myself that night, I created my bucket list of the most difficult things that I could possibly do in my lifetime. On that list, I wrote *rocket scientist,* amongst other challenging achievements. I thought about becoming a brain surgeon. However, the thought of seeing and working with human blood scares the living crap out me! So, I chose the next best profession – rocket scientist. That night, I fell asleep immediately after I wrote that list. Ironically, I forgot about the list until almost 15 years later through a circular chain of events.

I chose rocket scientist because of a first-grade, field trip. We visited the Jet Propulsion Laboratory (JPL) in Pasadena, Calif. For a child, it was like walking into an enormous candy store! I saw jet engines, runways, and the JPL Mission Control Room for space launches. I was fascinated by these high-tech computers, giant TV screens mounted against the walls, and countless red, plush chairs. I wanted to be like the men I imagined operated these rooms. I daydreamed a lot, and wished that I could help launch rockets. That memory still burns in my mind.

Fast forward 15 years later, when I graduated from college with my bachelor's of science in statistical mathematics: I was completely clueless about my next step. The truth is, I was afraid of not having health insurance after graduation. Low-cost, student medical care was a thing of the past. Plus, I knew that I wanted to attend graduate school. However, my preferred graduate schools didn't accept me. I had little income, and no money for graduate school. I was worried beyond belief.

So, I aimed for finding a full-time job, where I could receive medical benefits. My next priority was to gain a position that used statistics. Even though I received high university math scores, secretly, I had no clue about what I was *ACTUALLY* computing. As a result, I thought that if I could obtain a job using math and statistics, finally I could understand the concept behind each calculation. Perhaps, I could see math as real. Never did I consciously aim to be a rocket scientist. I still couldn't *REALLY* imagine myself as a rocket scientist.

After a two-month job search, I was hired by The Boeing Company. To my disappointment, I was not employed as a rocket scientist. Rather, I was hired to be a Quality Analyst. In the analyst role, I was to predict the cost of repairing, manufacturing and replacing the mechanical hardware. Truthfully, after the second day, I knew this role would be as boring as watching paint dry on a humid day. I wanted to quit. At the same time, I wanted to learn. So, I decided to challenge myself more by enrolling in an Applied Mathematics graduate program at CSUN, while still working full time. (I know I was insane.) After delving deeply into statistical math in undergraduate studies, I did not want my brain to atrophy in what I considered an unchallenging position. Plus, the company paid for 100 percent of my educational costs. Hence, I took advantage of the company's reimbursement program.

Steady Curiosity Opened My Doors

While attending graduate school, I studied courses such as Complex Variable Analysis, Stochastic Processes, Marchov Chains, Chaos Theory, and Multivariable Real Analysis. Yes, I had reached a milestone learning math calculations, considering my humble beginnings doubling oatmeal cookie recipes! Each class required a couple of days to solve one math solution. Some parts confused me. I needed help. Consequently, I asked around at Boeing, seeking a fellow statistician. I was directed to speak this "mysterious man" named, Zhao. A disclaimer also came with that advice, "Zhao's a genius! But English is his second language, so you might be able to understand about every eighth word." Dr. Zhaofeng Huang was originally from China. I was also told not to feel bad if I didn't understand him.

One day after work, I went to Zhao's office and met him. He spoke with a heavy Mandarin accent. I, on the other hand, left his office feeling very good. I completely understood him! Most students in my graduate school were from China. Zhao often worked late, because he had excessive work with no help. There were only two statisticians on the Boeing campus in Woodland Hills, Calif. No one else could help them calculate launch-related propulsion problems. Oftentimes, Zhao was forced to spend late-night hours finishing jobs.

A light bulb flashed! If I could help Zhao, then I could see how math was applied in reality. I explained to Zhao that I typically finished my work early. I also shared my background in mathematics. I told him that I was in graduate school, and that I loved math. I also admitted that I required some mentoring to gain the best solutions. I proceeded to ask him about my graduate math questions. Zhao became my mentor. By the way, he was a true rocket scientist.

After that day, I visited his office every day. Miraculously, I did something that no sane corporate person does: I volunteered to help Zhao after work for no pay. But, there was one major issue. I had no idea what the numbers represented. Previously, I had physics and engineering courses in undergraduate school, but I never learned *THIS* math. I witnessed this fact: math is a tool to represent something that happens in *real* life. This meant that I had to learn rocket science on-the-job. Talk about "on-the-job training."

Each time I received numbers from Zhao, I took the opportunity to introduce myself to the engineers who knew the numbers' origins. All were shocked when they saw this young, sexy, exotic, French-Indigenous/African-American female, who was curious about rocket engines. But, gaining information from them was like pulling teeth. Some were distracted. Many men were literally in awe by my striking facial features and shapely appearance. Sometimes, I waved my hand in front of their faces to get their attention. Others were annoyed. Plenty of males did everything they could to make me feel like I was interrupting their day. To these men, I was viewed as an inquisitive person that knew very little about rocket science. In any case, the majority did not feel obligated to share their time. Interestingly, they saw my physical characteristics before anything else.

Despite some of the engineers' opinions, I treated the situation like a game. I knew far more about survival than they ever would. I had a degree from *Hard Knocks University*. Many would never experience this type of education. The more information I gathered: the greater points I accumulated to win my psychological game. Each point represents a victory for my perseverance. Though, I possessed another motive. Every time that I interacted with the engineers, my goal was to make them feel comfortable around me. I was determined to prove how humans have far more similarities than differences, and can solve problems together.

This process allowed me to gather invaluable information about science and me. I learned about fluid flow, combustion thermodynamics, bending moments, fatigue cracks, thermal expansion, non-destructive testing, welding processes and failure modes. I also learned about golf, banking, interest rates, home purchases, tennis courts, sports and cars. As you might guess, I became well-versed on the subjects of steak houses, along with my co-workers' single sons and grandsons. My previous mundane role turned into an exciting position, where I became mentally stimulated, and rather entertained.

After months of working with Zhao and my original team, I received the *1999 Boeing Growth and Innovation Award*. I received this award because one of my quality reports helped save the company $100,000 in costs. A year passed. Then, I finally received the recognition I was awaiting. In 2000, management informed me that I would start working with Zhao and his team. I was told that my math skills were best used as a rocket scientist! My role was to calculate the probability of the catastrophic explosion for each of the Space Shuttles' three Main Engines. At the time, we had four Space Shuttles; *Discovery, Endeavour, Atlantis* and *Columbia*. My job was to foresee and mitigate potential explosions on the launch pad: during start or within flight.

When first hearing of this *breathtaking* assignment, my mouth dropped. I had no idea that I would *actually* be a rocket scientist. My childhood dreams had officially come true! After three years on Zhao's team, I received the *2003 Black Engineer of the Year Modern Day Technology Leader Award*, which was recognized for my mathematical analysis that helped secure $55 million in new engine contracts. The following year, I received *The 2004 Boeing Company Professional Excellence Award* for my innovative contributions to the company. These awards were a sign that I needed to continue making a significant difference in science, as well as breaking down stereotypes.

Mission Control Room Stories Exposed

A loud NASA voice counts down, "Ten. Nine. Eight. Seven. Six. Five. Four. Three. Two. One. Lift, off!"

"Pressure is good. Vibrations are nominal. No hydrogen leaks. We are good. Valve timing is good. Liquid hydrogen flow … unobstructed." I reported with authority. Now, I was supporting Mission Control. Large, mini-size movie screens hovered above our heads. I sat in this room with excitement, amazement and pure exhaustion. Per protocol, we were required to be in this dark room for over 12 hours before the launch, and sometimes overnight.

After almost a year of authorizing the system safety aspect of engine tests, and gaining the National Engineer of the Year Award, I was offered the opportunity to work in the control room. Here, we verified the Space Shuttles' three Main Engines safety during launches for Houston's NASA Control Center. This opportunity was the pinnacle of my rocket science career. Through years of demonstrating my ingenuity in reliable rocket propulsion, I was promoted to our company's control room, where we support Mission Control. I was trained by my co-worker, and second mentor, Dennis Early. He was a tall, older Caucasian male, who spoke slowly, but had a photographic memory. He could recite every engine, bolt, weld, and duct's location and name in a split second.

As the saying goes, "It takes a village to raise a child." And, in the case of the entire Space Shuttle Program, it took "several companies" to build the astounding spacecrafts. The Boeing Company made the vehicles. Solid rocket boosters were built by Thiokol. Boeing's Rocketdyne Division created the Space Shuttle Engines. Each company had their own room to monitor the launch. Accordingly, each respective program manager gives NASA the authorization to launch, if there were no problems within the launch preparation.

Within each preparation, we experienced pure excitement. At any moment, something could go wrong. A pump may explode. Valves may not open. Sensors may give false readings. Gaseous substances may catch fire. In these failure cases, we were expected to solve the issue within five to 10 minutes. NASA expected to launch the vehicle despite the temporary crisis. This meant that if a problem occurred, we, as an engineering team, were expected to solve it. The solution process was to; (1) identify the location of the problem; (2)

determine the methods for correction; and (3) implement a solution within a short segment of time. This correction was to ensure that the mission would not be endangered.

We are trained to watch events like a hawk. Failure is not an option. Success is now or never. As I look back, I understand the reason why NASA needs to launch without man-caused delays. It's estimated that each delay cost NASA over $5 million per day. This figure possibly has increased since I left. Hence, if we are to detect an issue that can possibly endanger the astronauts and the multi-million dollar cargo, we are not only expected to identify the problem, but also establish an action plan to resolve it. Sitting on the "hot seat," made our blood pump fast for 12 consecutive hours. Our job is to ensure *the mission*.

As I reminisce, the Mission Control room taught me the best lesson of all. We are expected to solve our problems quickly. In this way, our individual missions are not delayed or impacted. Once we are aware of our goals, we must be prepared to create an action plan to stay on track, despite our fears. Individual missions are so vitally important that we must take charge of separate mission control rooms – our brain. And in it, we must correct anything that may hinder the brain from functioning correctly. We must secure our own mission toward success.

The Epiphany before the Launch

"Why did you leave rocket science?" My college students continually ask.

Imagine eating your favorite food, every day of your life. Eventually, your taste buds will crave something different. Day-after-day, my engineering role began to taste the same. The role was extremely introverted and rather isolated. In fact, I was scheduled to have the exact same role for the next 20 years. Can you imagine that? Yet, my professional tastes and overall personality changed during my eight years at Boeing. Through my experiences, I became a more extroverted person. After I earned my master's degree in Applied Mathematics, I knew that I could make a more direct impact on the lives of others. I experienced a new calling that was far more influential and important than my current role.

In the days prior to one particular launch, 200-300 top engineers and scientists would gather in large rooms to discuss the technical criteria for launches. We reviewed every bolt and inch of the engine to verify its structural stability. We confirmed that the engine could endure excessive loads during the launch. We reviewed material integrity, manufacturing processes, engine timing and start sequences. And yes, we checked O-rings! We reviewed every factor leading to a successful launch. I personally presented every engine-related factor that could make the vehicle explode like a firecracker on the Fourth of July.

At the beginning of every meeting, *International Traffic-in-Arms Regulations* (ITAR) were announced. This ITAR meant that every non-U.S. citizen had to leave the room, so our nation's technical security could be maintained. Only authorized people could have access to this information. Some of our nation's most secret satellites were delivered through the Space Shuttle. The technical information was so secret that foreign governments sought to recruit spies to expose these U.S. technology secrets. *Later, I found out that I had been working with a spy...*

During the first couple of years of this ITAR announcement, a few people would leave. The next year, twice as many exited. Years passed, and only a few U.S. citizens remained. More and more top engineers were retiring. I began to calculate that in a few years, almost all the remaining engineers would retire.

My epiphany moment came:

- Who will perform this scientific work 20 years from now?

- How will the government maintain its technical security?

- Who will create U.S. inventions if no one is in the math and science fields?

- Will the U.S. become a Third World country in 50 years if we do not have Americans that can solve these technical problems?

These questions began my quest to change our U.S. future. More importantly, I realized that I would have a difficult time raising these concerns to the world within a 5' x 5' cubical doing math calculations 99.99 percent of my time. My tastes and my own mission had changed. I was now destined to expose math and science to the world! However, I was clueless on my next steps. Time passed as I waited to develop a feasible plan.

Rocket Scientist to Federal Banker

In life, we will always have tests. School exams are actually the easiest, because life tests are more challenging. Sometimes failure is a setback in one direction, but serves as a springboard that bounces toward success in a completely new direction. I have another motto: *"If I fall, then let me fall forward."*

When a person is brilliant, he or she may receive offers from multiple organizations. And if he or she is wise, that person will make 1+1 equal more than 2. In my case, I received an offer to be a project manager in a federal banking organization. It was an offer that I could not refuse.

"If I fall, then let me fall forward."

I was tasked to create a math model and strategy to prevent a $10 billion dollar currency shortage within all U.S. banking institutions. After much internal debate, I decided to accept the role. I aimed for expanding my management and math skills in a new direction.

I accepted the offer and provided a math-based solution plan that would strategically prevent this currency shortage. Unfortunately, I discovered that I was not a banker at-heart. I sat in a cubical with no regular human interaction. I am not sure if you have gathered this fact yet, but I am quite personable. At the time, I was complacent and accepted the situation. Later, I had several conversations with entry-level bankers across the U.S., who also needed help in math. Upon speaking with them, I began to re-evaluate my position. I saw that Olympia LePoint was supposed to assist more people directly instead of indirectly.

> *"Math is needed everywhere to make decisions."*

My banking experience led me to remember my original mission. I desired to share math with the world. I was a person who loved touching and healing people by teaching them through creative solutions and science. If I continued in this six-figure role without following my calling, I would accept second-best in exchange for money for the rest of my life. A "banker" was not the existence that I wrote on my bucket list when I was 8-years-old.

So again, I did what no sane corporate person does. I quit without a backup plan. However, my resignation gave me my ultimate epiphany yet: *"Math is needed everywhere to make decisions."* I have never regretted the decision to leave either company. For, I passed my real-life exam. Through these experiences, I catapulted back on my influential path. Hence, you are reading this book now. Nevertheless, with any choice, there is a trade-off. I fell, but I fell forward. The trick is to use math concepts to gain the desired set of outcomes – so we are always rising forward toward our goals.

DO WE REALLY NEED MATH?

CHAPTER 2

DO WE REALLY NEED MATH?

"When will you EVER use math?" I asked myself this question too.

We are experiencing an exciting time of science exploration. During the writing of this book, Mars rover, *Curiosity*, landed on Mars, Aug. 6, 2012 *(http://www.wptv.com/dpp/news/science_tech/mars-rover-curiosity-landing-landing-photos-videos-human-voice-makes-giant-leap-in-space-thanks-to-nasa)*. Everyone on Earth is excited, similar to when U.S. Astronaut Neil Armstrong first stepped on the moon in 1969. Many courageous men and women in NASA found the math and science tools to make this great Mars *Curiosity* expedition possible. Even non-scientists are excited about this new era of space technology.

With the help of Grammy-winning producer-songwriter and *Black Eyed Peas* artist, *will.i.am*, the planet Mars is already listening to its first "Earthly" song, "Reach for the Stars" *(http//:abcnews.go.com.m/blogEntry?id=17107122)*. will.i.am is daring to inspire the next generation by giving back to his Boyle Heights community, as well as encouraging the S.T.E.A.M. or science, technology, engineering, arts and mathematics education program!

It took eight minutes to completely transmit this electronic file from Earth to Mars! When I asked my students to calculate the distance between Earth and Mars using this eight-minute fact, they were puzzled. My students hadn't realized that math represents real-life events. I was about their same age when I recognized how math was used. I was an 18-year-old freshman in college.

I decided to count the number of times that I actually used math. My results were fascinating. In the morning, I estimated the average time to dress. When traveling to school, I gauged my safety by watching a car's speed and distance. While in class, I estimated the score I needed to earn an "A." At home, I used math to double baking recipes. As I ate, I optimized the consumption of protein to maintain my blood sugar level. I calculated tips for waiters. I even began predicting future events based on past occurrences. Countless people are unaware of how often we work with math daily. Would you believe the process of counting almost became an obsession? I found myself constantly counting...

In this chapter, you will see how people were swindled out of billions simply because they couldn't read financial statements. You'll also see how math is applied in rocket science, and will be able to determine how good (or bad) you are with basic math. Once you see how we've been using math over the centuries, then you'll understand the importance of overcoming mathaphobia®.

Math and money go hand-in-hand. I have always wanted multi-millions in my bank account. I innocently believed that if someone invested money wisely, their earnings would grow for years. I had this idealistic thought until I discovered a scandal that rocked Wall Street, where millions were stolen from CEOs and affluent individuals: simply because of one basic inadequacy. None of his Wall Street clients had the ability to comprehend math computations in their financial statements. Would you have spotted this scam if you were his clients?

In 1960, a "financial investor" founded the **Wall Street Firm** and decided to secure investments from many well-known, multi-millionaire and billionaire clients. These clients were individual and corporate charities. Instead of returning the profits to these investors through actual stock market earnings, this con artist decided to create a fraudulent investment scheme. The scheme gave payment to a few unrelated people. This man used what is frequently referred to as a *Ponzi Scheme* named after a 1910 con artist named Charles Ponzi.

From 1996 to 2008, Bernie Madoff, 73, followed a similar scheme that led to "the largest fraud in human history," according to Preet Bharara, U.S. Attorney for the Southern District of New York. Under the Bernard L. Madoff Investment Securities LLC, Bernard, his brother Peter, and several other "cohorts" carried out a multi-billion-dollar theft! For years, the individuals and charities had no clue that their money was being stolen, although the clients had mathematical warning signs. All funds returned consistent gains of more than 10 percent every year, despite the mathematical impossibility. The clients were only charged commission fees, instead of the normal commission fees plus the profit percentage fee. This meant that multi-million dollar earnings were mysteriously unaccounted for. What's more, the clients received earnings figures on letterhead, versus receiving an official actual stock and profit report performance on graph paper. Unfortunately, the clients suffered from innumeracy, and didn't realize they were being duped by their investor. At the same time, the clients put their whole trust into their investor and didn't know what they were supposed to receive. Thus, the clients failed to see that the math did not add up.

Concerns about Madoff's investment firm business began as early as 1999. A financial analyst, Harry Markopolos, informed the U.S. Securities and Exchange Commission (SEC) that he believed it was legally and ***mathematically*** impossible to achieve such elevated earnings. He wrote the paper, *The World's Largest Hedge Fund Is a Fraud*. Despite his efforts, Mr. Markopolos was ignored by the Boston and New York SEC. He presented further mathematical evidence. Unfortunately, SEC officials did not heed to Markopolos' warnings despite receiving mathematical evidence. With this temporary setback, Mr. Markopolos and his team continuously alerted the government, the industry, and the press about the fraud for more than 10 years. There appeared to be no hope in sight.

However in 2009, informants came forth. Bernie's own sons, Mark and Andrew reported their father to federal authorities. According to F.B.I. documents, Mark and Andrew met with their father to question his plans to distribute "hundreds of millions of dollars" in bonuses to employees, months ahead of schedule. This led to a private meeting, where he confessed to his sons about spearheading a giant Ponzi scheme that was crumbling. Mark was a licensed broker at his father's firm and many of Mark's childhood friends lost their savings behind the Madoff Ponzi scheme. Nonetheless, Bernie confessed and the trial was forfeited. He was ordered to serve to the maximum 150 years in prison for nearly five decades of fraud. This scheme stands as the largest financial investor fraud totaling $64.8 billion. Peter, 66, the chief compliance officer and senior managing director, plead guilty to, among other things conspiracy to; commit securities fraud, tax fraud, and mail fraud. Peter was ordered to forfeit several homes, a Ferari, and more than $10 million in cash securities, which will go toward compensating the victims of Bernie Madoff's Ponzi scheme. So far, the prosecutors have only found $21 billion of the loss, and $43.8 billion is still missing.

Sadly, on Dec. 10, 2011, Mark was found dead of an apparent suicide. It was on the second anniversary of his father's arrest. He hadn't spoken to his father since the day of the arrest, and had left several suicide notes to family and friends. Many say Mark had also been the victim of numerous death threats, and pending lawsuits as more than $60 billion has yet to be recovered to victims.

> *"In the Madoff Ponzi scheme, math-illiteracy affected the most financially powerful CEOs and SEC officials."*

Now, you may ask, "What prevents another Madoff Ponzi scheme from occurring?" The answer is ***nothing***. Such schemes occur fairly often, but not at that great level. Numerous still do not know how to read, interpret, nor use math with respect to financial statements.

It is my belief that either the millionaire clients thought that the math was too confusing, or they were part of the scheme. Either way, people suffering from innumeracy lost the most.

"In the Madoff Ponzi scheme, math-illiteracy affected the most financially powerful CEOs and SEC officials." I am certain that when we know math, and its cause-and-effect logic, we can ward off any predators and poor choices.

Math is a tool to represent the past, present, and predict the future. In using math, we thrive as humanity. Without math, we stand at the mercy of man's self-induced "chaos." Our historical development, current technology, and future economy depend on our individual efforts to read and use math.

Math Is Nothing New

For more than 14 billion years, the universe has existed. Some 3,000 of those years, humans applied math to represent and predict natural occurrences within time and space. Before the Greek alphabet, math was represented in pictures. And, we still use pictures to share the "math story." Throughout time, math has been translated into various languages

For example, in 2700 B.C., Egyptians used hieroglyphics to demonstrate time, medical procedures, star rotations and continent locations. Later, Greek and Roman empires translated these hieroglyphics into their alphabets. The Mayans devised a counting system that represented very large numbers using *only* three symbols. Native Americans also created advanced probability calculations for gambling and meteorology. They also created extremely-advanced, mathematical codes for smoke signals. And during World War II, the U.S. military joined forces with the Navajo tribes to create an intricate communication system. The U.S. Marines recruited Navajo men as "Navajo Code Talkers" to transmit the **Navajo Codes** from 1942-1945. The code was a combination of military language, Navajo words and letters of the alphabet.

> *"Throughout time, math has been translated into various languages."*

Finally, the Australian Aborigines also made predictions about seasons and weather. Now in our 21st Century, we use computers to tell us the "pictures."

Native Americans used mathematical smoke signals to communicate over teepees across tribal communities, as seen next to Olympia LePoint.

Currently, in rocket science, we capitalize on computers and computing software to tell us about the math story in the rocket engine's operation. Sometimes, we use hand calculations. Oftentimes, we use computer-based simulations and math-based computer programs. All calculations help us to identify the potential areas for explosion in the Space Shuttle Main Engines. We refer to a desired explosion as *engine combustion*.

Picture source: NASA

A Space Shuttle launch controlled combustion

Picture source: NASA

Space Shuttle Main Engines controlled combustion in flight.

For us rocket scientists, computer technology can be both a blessing and a curse. While at The Boeing Company, there were times when we simply input numbers into a software program, and the answers easily rolled out. On other occasions when on deadline, data was quickly input and our computers slowly provided outrageous conclusions. In panic mode, we'd pull out our pencils yelling, "Quick! Quick! Let's do it ourselves." As a result, we scratched our heads and performed the calculations by hand.

Still true today, rocket scientists have to be precise! Numerous engineers depend on our math calculations to design, redesign and test their delicate hardware components. Each engineer possesses his or her own expertise. We, as a program team, combine our specialties together to build the engine.

There are experts in heat transfer within the engine, while others specialize on steel movements caused from high levels of force. Some scientists are skilled in electronic wiring. Many are top authorities on the manufacturing processes. All rocket scientists have specific math calculations, which cohesively work together to build each Space Shuttle engine.

My expertise was to identify the interactions that could cause each engine to fail. I used cause-and-effect logic and statistics to show the probability of an undesired combustion. Bluntly, I predicted how each engine could explode. Needless to say, all the engineers depended on my intelligence to help show the future. Each time, the engineers needed our calculated probabilities.

The Space Shuttle Main Engine During Testing

Picture source: NASA

> *"All rocket scientists have specific math calculations, which cohesively work together to build each Space Shuttle engine."*

I had another significant responsibility. I authorized the safety plans for engine tests. I verified that the hardware was intact, appropriate sensors read temperatures and pressures correctly, along with ensuring that the test stand's safety controls were in place. This effort was to ensure no test engineer would die if an accident did occur.

Thankfully, our math efforts paid off! We successfully launched 28 Space Shuttles while I was at the company from 1998 to 2007 *(ref: http:// en.wikipedia.org/wiki/Space_Shuttle).*

As I write this book, NASA's U.S. Space Shuttle *Endeavour* – the aircraft I helped launch into space – took its final aerial victory lap over numerous Southern California sites. The shuttle arrived in California on Friday, Sept. 21, 2012, after a cross-country farewell tour. Since NASA's 30-year Space Shuttle Program retired in 2011, this is NASA's final Space Shuttle ferry flight across The United States. A modified Boeing 747 carried the shuttle as it flew over *Los Angeles City Hall, The California Science Center, Disneyland , The Griffith Observatory, Universal Studios, NASA's Jet Propulsion Laboratory, California Institute of Technology*, as well as Malibu, downtown Los Angeles and other locations *(ref: http://latimesblogs.latimes.com/lanow/2012/09/ space-shuttle-endeavour-flyover-and-landing-viewing-tips.html).*

Finally, the *Endeavour* touched down at Los Angeles International Airport, making its last landing after a three-decade career in space, and bringing an end to NASA's Space Shuttle Program. I am saddened because NASA's human spaceflight program has officially ended. However, I am now thrilled!

Picture source: LA Times

Today, I tracked *Endeavour* by watching everyone's picture posts on popular Social Media sites. I am happy that others share the same excitement that I had, while working for NASA's space program. And, when I personally saw *Endeavour* fly above, I felt like a kid again, similar to when I visited the *Jet Propulsion Laboratory's Mission Control Room* when I was 6 years-old. I am reminded that it was a privilege to work as a rocket scientist for NASA's space programs. While there, I was inspired to solve big problems as I worked as a rocket scientist. I cherished the opportunity to build and launch vehicles that fly like planes, glide like birds and shoot like rockets!

Now, my mission is to help other men and women; young and old; rich and poor; privileged and underprivileged to soar to new heights like I reached, while working to launch *Endeavour.* I am ecstatic, because a new generation of innovators – like the little 6-year-old Olympia at *JPL* – are watching this experience, too! They will be our next generation's great solution-finders! Today has been a wonderful, historic day, which has inspired more future "Olympias" to pursue the unthinkable.

Picture source: www.space.com

You Can Do It, Too!

Now, you may never desire to be a rocket scientist. In fact, you may not desire to be a mathematician either. But, if the math illiteracy phenomenon scares you, there is one thing that you can do. You and your loved ones can become math-literate and end your math fear. Change your future. Continue reading, and take a step toward gaining the tools you need to make your courage override your fear. First, let us start with a basic math questionnaire.

How Good Are You with Basic Math?

Answer the following five (5) for the next ten (10) minutes. During each question, observe your thoughts. See how many questions that you can answer within 10 minutes without a calculator. Set your timer. The answers are on www.Mathaphobia.com under *Mathaphobia® Book Answers.*

1. Convert 5/8 to a decimal and a percentage.

2. Arthur goes to the bank to deposit cash. He has 35 bills totaling $420. The money was either $10-bills or $20-bills. How many bills of each does he have?

3. What is the difference between a greatest common factor, a least common multiple, and a least common denominator? Which process is used when "clearing the fraction?"

4. Two numbers have a total of 83 and a difference of 17. Find the two numbers.

5. If in the first year, Paul's car value is $15,000, and in the fifth year the value is $10,000. The car depreciated according to the linear model y = -$1250x + $16,250. If he wants to sell the car for $9,000 in the eighth year of ownership, is he asking too much? What should he request? Based on the formula, what was the car originally worth? (Disregard mileage in your calculations.)

TIME!!!

Okay. Your 10 minutes are up. Did you feel comfortable analytically figuring out the math set-up, the formulas and the solutions? Congratulations! If you felt uncomfortable, or if you had 5 questions or less correct within 10 minutes, keep reading. This book is definitely for you.

Do You "Math–Sabotage"
Yourself?

CHAPTER 3

DO YOU "MATH – SABOTAGE" YOURSELF?

Ready for the journey? In this chapter, you will learn about the detrimental affects of mathaphobia®, understand the characters that self-sabotage themselves because they suffer from mathaphobia®, and determine if you, indeed have mathaphobia®. But, before I reveal more about mathaphobia®, I will give you this analogy. You just won a getaway vacation for one person to a resort! A shuttle bus is scheduled to arrive at 8:45 a.m. to transport you and other winners. Food, travel, lodging, and excursions are all-inclusive. Once there, a limousine will take care of your additional travel. Truly, you want to experience this destination. However, you secretly doubt that you should enjoy this experience alone. Perhaps, you want your children or your partner to join you. Possibly, you think that your best friend should accompany you. In any case, you hesitate on the travel. As a result, you unconsciously sabotage your vacation prize by committing irrational actions. Here are a few unreasonable scenarios:

- **Self-Sabotage Case 1:** Swamped with immediate concerns, you wait until the last minute to pack. You stay up all night, and it's now 8:45 a.m. when you board the shuttle bus – beat with exhaustion. Too tired to maintain a conversation with other passengers, you start falling asleep. You doze in-and-out of consciousness. Periodically, you see the travel route, but are so weary that you're completely disoriented. The bus arrives back at your original destination, and no one awakens you. Sadly, you were asleep and missed the opportunity to get off the bus. The shuttle then returns to your original pick-up location. But, you're still on the bus! You remain forever livid with yourself for missing a great opportunity. Years later, you're still beating yourself up for falling asleep on the shuttle bus.

- **Self-Sabotage Case 2:** You arrive at the bus stop four times in total; (1) To check out the bus stop location; (2) Arrive two hours earlier to determine the traffic flow; (3) Visit 30-minutes before the scheduled pick up. At this time, you have an urge to be more prepared. Thus, you go back to your car to get your umbrella and; (4) you return and realize the bus came, and left without you. You are irate at yourself for returning to the car, and failing to know the EXACT bus schedule.

- **Self-Sabotage Case 3:** Due to a weak knee, you have difficulty traveling to the shuttle pick-up location. Consequently, you arrive late near the bus stop, finding people onboard with an apprehensive bus driver waiting. While slowly walking to the shuttle stop, you fail to indicate your physical restriction to the bus driver. Looking crossly at you, the driver motions for you to "hurry," and board the bus. However, you're prideful, and neglect to admit you can't run because of your knee. The driver thinks that you are too slow, and is delaying the bus schedule! Thus, the bus driver closes his doors, leaving you at the shuttle stop! You are not physically able to quickly run after the shuttle. Due to your physical challenges, you are angry at yourself for not being physically capable of experiencing a great opportunity.

- **Self-Sabotage Case 4:** You happily board the shuttle, sit four rows back from the driver, and settle down for a bumpy ride. Anxious to start this well-deserved getaway, you become frustrated when you think the driver is slowly cruising. You start "back seat driving," telling him how to take certain routes to quickly arrive at the location. Assuming you know a better route, but you don't, the driver becomes highly-annoyed. Your blood is boiling, and you can't endure the driver's obvious disregard to your suggestions! To save from further embarrassment, you demand to get off the bus. He finally lets you and your luggage off. Yet, you're a mile away from the destination, walking in the scorching sun! Now, you are furious at the bus driver, because he didn't take your suggestions.

Each winner sabotaged his or her prize of experiencing a new adventure. In Math, we do the same. Oftentimes, we hinder our self-development by behaving irrationally.

Such illogical actions stem from an unconscious fear. If the fear is not removed, our thoughts and actions can be warped. The phobia traps us in a repetitive cycle, void of growth and excitement. Once we understand fear and its effects on our thoughts, we can minimize our self-sabotaging actions. More importantly, we can maximize the opportunity to accept our future gifts. First, we must understand fear and its subconscious contribution to stagnation.

The Effects of Fear Are Real

Are you afraid of spiders (arachnophobia)? Heights (acrophobia)? Snakes (ophidiophobia)? Confined spaces (claustrophobia)? Going to the dentist (dental phobia)? Failing math (mathaphobia®)? What happens when you think of these fears? Does your heart race? Do you think of ways to avoid such terror? Do you cringe and want to do *ANYTHING* to get away?

"Oftentimes, we hinder our self-development by behaving irrationally."

If any answer is yes, then you have now triggered fear in your brain. According to Wikipedia, fear is an emotional response to threats and danger. It's also a basic survival mechanism that occurs in response to a specific stimulus, such as pain or the threat of pain. As mentioned above, the stimulus may be a spider, dentist, snake, tall heights, or even math.

Your phobia triggers the "fight-or-flight response," also called the "fight-or-flight-or-freeze response," in the reptilian part of your brain. In 1929, Walter Cannon, an American physiologist and professor at Harvard Medical School, first discovered this phenomenon when he tested animals. Cannon found that a mammals' fight-or-flee response is also stimulated when there is fear or threat of danger. During such scenarios, the brain releases epinephrine (adrenaline) and norepinephrine from the adrenal glands. Next, the body halts or slows down various normal, daily bodily processes such as the digestive system. Arousal responses and creative thoughts also adapt in order to survive the stressful situation. When a person remembers a past trauma, stress responses can sometimes go haywire.

Can you remember your first experience with a horrible teacher? For me, my educational trauma began in fourth grade. I had a horrible teacher, who wore immaculate clothing and way too much perfume. "Mrs. Subtraction" found pride in degrading every student who answered incorrectly. In my fourth-grade jail, (known as her class), I erroneously assumed that I had to be perfect to gain appreciation. Otherwise, I would endure her verbal abuse. I feared making a mistake. Consequently, my future performance-related stress was sometimes associated with my fourth-grade trauma.

During the time I was having extreme fear attacks, my body took on severe panic reactions. I started placing my hands near my face. My heart started racing, and I would sweat profusely whenever I didn't know an answer. When asked to convey my comprehension, I would freeze. Gripping my pencils tightly, I would bite my lip and nails.

The effects of fearing math are universal as well. Now as a math educator, I witness the same terror in students at the start of every semester. Students place their hands on, or around their faces. Some completely turn away from the board. I see students grip pencils tightly, or bite their lips, nails, heavily breathe or hold their breath. Others shake in hesitation as they write an incorrect step. When I see this, I know there is more work ahead of me. Such students can't hear nor understand anything that I say until this terror is eliminated. Unfortunately for many, their fear is so gripping that it stops their daily bodily functions. It is my belief that if this fear is not resolved within a person's brain within seven days, it transforms into a phobia.

What is Mathaphobia® Exactly?

Studies have shown that if we do the same action for seven consecutive days, the action turns into a habit. When math apprehension is unresolved, the fear turns into the habitual, contagious mathaphobia® mental virus. Math has not changed: however individual thoughts toward math have transformed. Earlier in Chapter One, the basic Mathaphobia® definition was given as; *the fear and contagious mind virus that blocks the brain from math literacy, basic math calculations and analytical problem solving*. Now, we will understand mathaphobia®'s specific affects within the body and mind.

In more detail, mathaphobia® is a mental, self-imposed fear, causing the inability to deal comfortably with the fundamental notions of numbers, chance and cause-and-effect relationships. Mathaphobia® creates an unnatural, illogical functioning of the brain with physical and emotional effects. Your heart tends to beat fast, palms sweat, and your brain constantly fails to comprehend presented information. In some instances, mathaphobia® can cause as much difficulty as a learning disability and mild, self-imposed retardation.

Mathaphobia® is linked to the amygdala, an area of the brain located behind the pituitary gland in the limbic system. The amygdala secretes hormones that control pain, fear and aggression. The amygdala also aids in the interpretation of emotion in the facial expressions of others. For example, if you see someone jump in fear, naturally you become afraid as well.

Location of the Amygdala in the Reptilian Brian

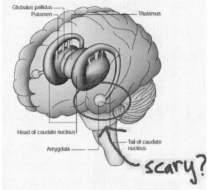

This area correspond to the onset of fear and pain.
Source: Society for Neuroscience

When the fear or aggression response is triggered, the amygdala releases hormones into the body to place the human body into an "alert" state, in which they are ready to move, run, fight, or freeze. With the fear, the body continues to secrete the 'alarm' hormones, even after the person rationalizes the situation they are experiencing. The amygdala stops releasing hormones when creative thought is activated.

Rear View through Brain Imaging

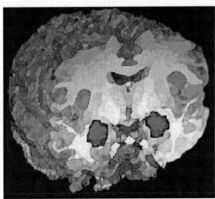

Source: NIMH Clinical Brain Disorders Branch 3-D MRI rendering of a brain with fMRI activation of the amygdala highlighted in red

Unlike other phobias, mathaphobia® gradually and constantly releases hormones during exposure to math, or the thought of math-related subjects.

> *"As with other phobias, mathaphobia® cuts off the physical blood flow to the brain's frontal lobes."*

Mathaphobia® may produce the following continuous signs of infection:

- A persistent, irrational fear of a specific math activity or situation.

- An immediate response of uncontrollable anxiety when exposed to math.

- A compelling desire to avoid math and taking unusual measures to stay away from math topics.

- An impaired ability to function in the presence of basic math calculations.

- Extreme math terror after a triggered event.

- When facing math, an experience of panicky feelings, such as sweating, rapid heartbeat, avoidance behavior, difficulty breathing and intense anxiety.

- In some cases, anxious feelings when merely anticipating an encounter with math.

As with other phobias, mathaphobia® cuts off the physical blood flow to the brain's frontal lobes.

The frontal lobes have several functions. They are responsible for planning, "executive functioning," abstract problem solving, staying on task, paying attention, and innovative thought. In addition, the frontal lobes are responsible for controlling impulsivity, as well as recognize future consequences resulting from current actions. The frontal lobes help choose between good and bad actions, along with determining similarities and differences between things or events. As you can see, the frontal lobes are involved in higher mental functioning.

Location of the Frontal Brain Lobes

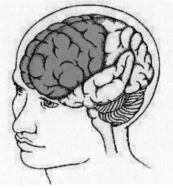

Source: The Neurobiology of a Zombie

Mathaphobia® activates the reptilian portion of the brain responsible for the "fight-or-flight" response. In the presence of mathaphobia®, all frontal lobe function stops. As a result, creative thinking – the pre-requisite brain energy required to transform math problems into solutions – is halted. This means that mathaphobia® and math problem solving cannot exist simultaneously in the brain. Specifically, math problems cannot be solved when mathaphobia® is present in an individual's brain.

Location of the Frontal Lobes and Reptilian Brain

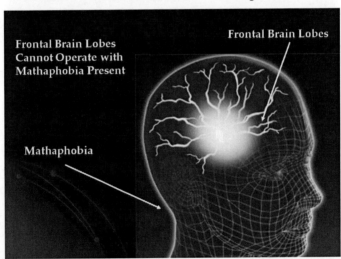

Source: Math Made Real Keynote, of Olympia's End Mathaphobia Now

If mathaphobia® resides and festers in the unconscious mind, then the conscious mind takes alternate actions, which do not require numbers nor cause-and-effect-related thought. For example, let's look at a cookie recipe. If you must double ¼ cup of sugar and didn't know it would equal a ½ cup of sugar (which you have in your cupboard), then you might go to the store and buy a pre-made cookie package. Instead of baking the cookies yourself, you feared that you didn't have enough ingredients at home. This is an example of when the frontal brain lobes are unable to create innovative thought, and rational reasoning. The good news is that mathaphobia® ends with a special recipe!

Types of Mathaphobia®

I have taught for more than a decade at colleges and tutored more than ten thousand students. Within this experience, I discovered that there are four categories of *Mathaphobia®*.

1) **Fear of Math Failure**
2) **Fear of Math Under-Achievement**
3) **Fear of Math Physical Learning Disability**
4) **Fear of "Math Control"**

Know Your Mathaphobia®

I will give you a quick-and-easy way to remember each category, by explaining the mathaphobia® person associated with each category.

A person suffering from mathaphobia® usually generates self-sabotaging feelings and actions toward mathematics. Such students are highly-intelligent, gifted and capable, but fear cripples their rational thinking. Here are the four (4) types of mathaphobic® people. Who are you?

"A person suffering from mathaphobia® usually generates self-sabotaging feelings and actions toward mathematics."

Quincy the Quitter

Quincy is a good student, yet he thinks that he is not as good as other students. *Quincy* tries hard to understand mathematics. But, he frequently thinks to himself, "I am not going to understand math." He has negative thoughts before he even sees his math assignments.

Quincy hates going to his math class because he feels lost. So he sits in the back of the class. He never asks the teacher questions, because he fears that the professor will make him feel stupid. Or, even worse, *Quincy's* afraid his teacher will make him LOOK stupid in front of others. When *Quincy* arrives home, he dreads opening his math book. He hates reading math questions that he doesn't understand. *Quincy* rationalizes in his mind that it's okay if he fails the class. He accepts that he will have to take math over, but he will still attend class to be consistent.

Whenever *Quincy* views math, he sees himself as a failure. He hates this feeling. When *Quincy* thinks of math, his heart begins to race: and he feels shame. A reoccurring feeling of neglect haunts him. Secretly, *Quincy* worries that he will be ignored and neglected when he tries to do something new. He thinks, "There is no one who can help me understand this math. I am all alone in this class."

Quincy blocks himself from learning math before he even tries.

Internal Thoughts & Feelings:

Quincy usually thinks to himself:

- I am never going to understand this math.

- I am not that smart.

- No one will want to help me to understand it.

- I fail at things in my life.

- I will never be successful, so why should I try.

Mathaphobia® 1: Math Failure (*Quincy*)

Within this mathaphobia®, the student is discouraged and no longer seeks help. Years pass before the student reaches out for assistance, and says, "I don't know where to start."

At an early age, this student may have been neglected or emotionally tormented by people in their immediate support circle. As a result, the individual feels as if he cannot obtain success in life's challenging areas. This includes math. The student feels like he will immediately fail on his first attempt. *Quincy* believes it's easier to not try at all. Avoidance behavior is falsely perceived as a protection from future failure. In turn, the student loses self-confidence in his own problem-solving abilities. The individual fails to constructively manage their growth and learning process.

Donna the Over-Doer

Donna does well in every class EXCEPT mathematics. *Donna* likes math, but she feels Math doesn't like her. She has been required to repeat math courses in the past. English, writing, music, drama and biology may be her fields of expertise. However math is not. Consequently, she believes that she will have to spend more time to understand math.

Donna does everything that she can do. She meets in a study group and memorizes terms. She goes to the teacher for help and makes flash cards. Although on tests, she spends unnecessary time trying to demonstrate her knowledge. She is confused with all the different types of formulas and examples. *Donna* becomes panicked, and never answers the questions correctly. She's so frustrated because she believes that excessive studying should lead to "A" scores. *Donna* doesn't understand the big picture or major concepts behind each calculation.

Her math tests are a reminder that she is a failure in life. Though in math, she feels like she is "never good enough." She believes that she can never "do things right." All of *Donna*'s other accomplishments are pushed to the back of her mind.

Internal Thoughts & Feelings:

Donna frequently thinks to herself:

- I can't memorize all the formulas. I have no idea what formula to use.

- I don't know what I am supposed to do.

- I am smarter in English, writing and other creative subjects.

- I have to do well in my life in order to be accepted.

- I only have to take the math requirement, and then I can stop.

Mathaphobia® 2: Math Under-Achiever (*Donna*)

A lack of mental organization characterizes this mathaphobia®. In this innumeracy-causing factor, the student has difficulty visualizing what the numbers represent: the "big picture is lost." This student is not accustomed to critically understanding why the particular mathematical computation was performed. This lack of "big picture thinking" creates difficulty in establishing the links between math situations and the numbers in which they represent. The student is not aware of what she is expected to investigate with respect to the major concept.

A student in this mathaphobia® category has no uniform method of organizing theorems, formulas and proofs. Therefore, she doesn't understand how, and when ideas overlap, and link together. The lack of organization restricts the comprehension of the big picture.

At an early age, this student may have felt abandoned by family members or people close to them. Thus, she feels that the more she does, the more approval she will gain to receive the desired attention. When her scores are not acceptable, she feels under-accomplished and develops a low self-esteem.

Samuel the Struggler

Samuel is one of the nicest people you will ever meet. He is personable and smart, as well as optimistic and charming. Teachers and students enjoy him. Yet, when *Samuel* starts to learn math, he becomes shy and ashamed of his self-perceived weakness. He struggles from any of these scenarios: he sometimes sees numbers and letters inverted; he has Attention Deficit Disorder; he cannot focus on the math and his minds starts to wander; he cannot read the board at the same time as he listens; or, he may not be able to write as he hears the lecture. *Samuel* looks at the test and does not understand any questions.

On tests, *Samuel* writes all over the paper. Yet, there appears to be no correlation between concepts and thought in his math work. He writes in the paper's corners, across the written questions, and on other pages. *Samuel*'s work appears highly-confusing, and he may write sentences on top of one another. When teachers view his work, they immediately give zero scores since his writing is illegible. *Samuel* thinks that all math teachers are disappointed in him, because they believe that he is not smart. Unfortunately, *Samuel* feels angry because he's unable to express himself.

He feels like a failure, because he knows that his mind does not work in the same way as his peers. As a result, he may drop the course because he is falling too far behind the other students.

Internal Thoughts & Feelings:

Samuel frequently thinks to himself:

- Wait! This math is going too fast.

- I feel so scatter-brained… I can't absorb what the teacher says.

- I have trouble focusing in math. I can't help but think of other things.

- I feel inferior to other people, because I am different.

- My math teacher thinks I am stupid, because she can't read my writing.

- I sometimes transpose numbers, and I write numbers in the wrong order.

Mathaphobia® 3: Physical Learning Disability (*Samuel*)

Visual and other sensory processing weakness is a condition that afflicts students with this mathaphobia®. This student fails to write a solution in a step-by-step, sequential pattern. A student in this category has difficulty sequencing concepts in math problems. He often has trouble completing solutions, which leads to even more confusion during their next attempt at solving problems. This unfinished state serves as a mental block.

On a student's paper, one might view a portion of an irrelevant formula or calculations in all corners of the paper, including a correct final answer without steps. Because there is not a clear logic from the student, he is accused of cheating or misunderstanding the problem. At an early age, these students may have been considered "slow" or "disabled" by their peers or educators. Sadly, such students often feel inferior to others while learning.

Crystal the Criticizer

Crystal is a student who likes to master everything. In fact, she's quite brilliant. She enjoys making events work to her advantage. *Crystal* loves helping people, and showing the methods for others' success.

However, she doesn't do well in mathematics. Consequently, she blames everyone else for her poor performance, and is obsessed with controlling the outcomes. She outwardly (or internally) disagrees with her teacher(s), because she feels that her way is best.

Crystal becomes so frustrated with her math teacher that she begins skipping class. She feels that she can learn math better from a tutor or herself. So she spends countless hours and money on outside tutoring. Or, she may try to learn the problems herself, and she learns an incorrect process.

Sometimes, she argues with the teacher about her scores. She feels that she can self-learn the information. But in reality, she incorrectly solves math problems. She is not open to learning a new way of doing math. For if she was, she would realize that it is okay to be a beginner again.

Internal Thoughts & Feelings:

Crystal frequently thinks to himself:

- I don't feel like being in this class.

- The teacher doesn't know what he's doing.

- I can do math better than the class explanation.

- I can learn the subject better from a tutor and not the teacher.

- There are very few people who I can rely on in my life.

- I want to do math my own way.

- I want full credit on my tests even though it's not the correct answer.

Mathaphobia® 4: "Math Control" (*Crystal*)

This mathaphobia® is characterized by the inability to be vulnerable. Math acceptance is letting go of past math failures and accepting the new aspects of math learning. Math acceptance is a skill that this student lacks. This student fails to yield to an instructor's step-by-step solution. She has difficulty accepting different math approaches to add to her repertoire. *Crystal* also does anything to avoid admitting ignorance. She goes overboard with hiring tutors to cover up her inability to solve the problem on her own. She often has difficulty respecting the math educator, and in turn, the instructor labels her as a "problem student."

At an early age, *Crystal* has been given adult responsibilities. Thus, she falsely believes that if she can control situations, she can avoid disappointments. She tries to control all aspects of her growth and self-development. This controlling behavior blocks exterior help.

My Mathaphobia® Self-Test. Who Am I?

Okay. Now, you know the mathaphobia® characters. Which character are you? As for me, I was *Quincy* the Quitter in high school. I never understood the material, and I frequently was lost. However, when I entered college, I became *Donna* the Over-Doer. I studied and memorized theorems without a strategy. We all have been one mathaphobia® character at one time or another within our lives.

But, don't give up! I have great news. You can take the mathaphobia® self-test to find out your phobia character. Are you *Quincy* the Quitter, *Donna* the Over-Doer, *Samuel* the Struggler, or *Crystal* the Criticizer? The test is subjective, so answer as honestly as possible. Remember, the questions are designed for you, and no one else.

For the following questions, select your response by the following criteria: 5 = strongly agree; 4 = sometimes agree; 3 = neutral; 2 = sometimes disagree; 1 = strongly disagree. The test is short.

---

1. I often feel as if my question doesn't count in math classes.   5 4 3 2 1

2. I am known to be good at many things, just not math.   5 4 3 2 1

3. At times, I see numbers backwards. Or, in ways others cannot see.   5 4 3 2 1

4. I sometimes think that I should know how to solve the math problems.   5 4 3 2 1

5. I tend to avoid asking questions: for it is typically not important.   5 4 3 2 1

6. Math is just not my subject.   5 4 3 2 1

---

7.   Many times, I have difficulty working with others when doing

math problems.   5 4 3 2 1

8.   I sometimes ignore the way in which teachers instruct class.   5 4 3 2 1

9.   I have anxiety starting math problems.   5 4 3 2 1

10. I am much better at communication and art, (than) not math.   5 4 3 2 1

11. I have difficulty seeing the board, hearing math lectures, and the teacher

goes too fast.   5 4 3 2 1

12. I have learned a more efficient way to solve the problems.   5 4 3 2 1

13. I typically do not do math homework.   5 4 3 2 1

14. I often do not understand why a teacher uses one approach over another.   5 4 3 2 1

15. Sometimes, I feel that I may have a learning disability.   5 4 3 2 1

16. I spend countless hours trying to self-teach math concepts.   5 4 3 2 1

17. I have failed a math class before and I have difficulty in math classes.   5 4 3 2 1

18. I either bite my nails, tap my pencil or feet, or fiddle with my belongings

when I receive a math test.   5 4 3 2 1

19. It takes me longer to do math problems.   5 4 3 2 1

20. I hate feeling stupid and not in control.   5 4 3 2 1

21. People say that I stop myself from excelling at problem-solving.   5 4 3 2 1

22. I get scared when thinking about my math homework. 5 4 3 2 1

23. My teachers have complained that they cannot read my math writing. 5 4 3 2 1

24. I have stopped attending class, because I cannot understand the material. 5 4 3 2 1

25. Few teachers pay attention to me and answer my questions. 5 4 3 2 1

26. I have received many honors and awards in non-math subjects. 5 4 3 2 1

27. Most of the time, I mentally cannot focus on math lectures. I zone out. 5 4 3 2 1

28. For incorrect scores, I have argued to receive full credit on math problems. 5 4 3 2 1

29. I have been frequently told that, "I never can do anything right." 5 4 3 2 1

30. When I know the reason why I would use certain formulas, I can understand math better. 5 4 3 2 1

31. Sometimes I cannot focus; I have been anemic or had a thyroid imbalance. 5 4 3 2 1

32. If I do not do tasks, no one else will get them done. 5 4 3 2 1

33. The thought of doing math scares me, and I do not want to think about it. 5 4 3 2 1

34. I usually spend countless hours studying math, but I still do not understand math. 5 4 3 2 1

35. When feeling stress, I stutter or sometimes have mental blocks in understanding math problems. 5 4 3 2 1

36. I can do math better than the class explanation. 5 4 3 2 1

37. I am always average at things when I attempt to do my best.   5 4 3 2 1

38. I have trouble understanding tutors' explanations and approaches.   5 4 3 2 1

39. The teachers cannot understand my writing or solutions.   5 4 3 2 1

40. The math tutors are more helpful than the teachers.   5 4 3 2 1

Self-Assessment Results

To find your results, you will add four separate groups of answers that represent the different characters.

• Phobia 1:

Add your numeric results together from questions:

#1, 5, 9, 13, 17, 21, 25, 29, 33, 37

• Phobia 2:

Add your numeric results together from questions:

#2, 6, 10, 14, 18, 22, 26, 30, 34, 38

• Phobia 3:

Add your numeric results together from questions:

#3, 7, 11, 15, 19, 23, 27, 31, 35, 39

• Phobia 4:

Add your numeric results together from questions:

#4, 8, 12, 16, 20, 24, 28, 32, 36, 40

VALUES OF ANSWERS

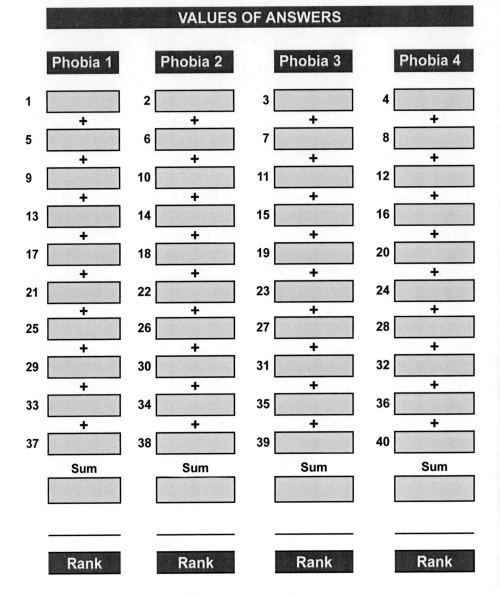

Now rank your results:

(Note: If you scored the same score for two groups, that's okay)

- **Phobia 1 represents the results for *Quincy* the Quitter.**
- **Phobia 2 represents the results for *Donna* the Over-Doer.**
- **Phobia 3 represents the results for *Samuel* the Struggler.**
- **Phobia 4 represents the results for *Crystal* the Criticizer.**

RANK	**SCORE**	**Mathaphobia® Person**
Highest score	= _____	_____
Second highest score	= _____	_____
Third highest score	= _____	_____
Fourth Highest score	= _____	_____

Your top-ranking phobia category score is your mathaphobic® character that is most similar to your current thinking. Empowered with this information, you now have the ability to end your mathaphobia®! If you are *Quincy* the Quitter, you can be David the Determined. *Donna* the Over-Doer can become Sarah the Strategist. *Samuel* the Struggler can transform into Ivan the Innovator. And, if you are *Crystal* the Criticizer, you can become Ellen the Explorer. You can achieve success by following a simple 3-step plan. Our next chapter will show you how.

OPEN YOUR HEART =
OPEN YOUR BRAIN

CHAPTER 4

OPEN YOUR HEART=
OPEN YOUR BRAIN

Math fear is not your fault. Yes. You heard me correctly. You are not to blame. Most likely, you feel rather overwhelmed now. You want to understand math, and you want a successful outcome. However, every time you think about failing math, you have a large knot in the bottom of your stomach. You feel doomed if you do, and doomed if you do not. You want to make a change in your life, but all your thoughts place you back into the cycle of math terror. The fear replays itself out through many areas in your life. You don't want to make it your crutch: although, you do not have any other tools except for fear. You want help, but you have no idea where to go, and who to talk to for assistance.

Your frustration seems maddening, because you need a way to break free from your phobia. Ideally, you'd like to operate in confidence. But, you are doomed to repeat the frustrating cycle of math illiteracy. You need a clear understanding of the root of your frustration, the tools for reprogramming your mind, and a solid action plan. Only you can break your pattern!

There is hope. Within this chapter, you will receive enlightenment. You will discover the true origins of mathaphobia®, the effectiveness of mind reprogramming, and the 3-step action plan to remove your fear for good. If you can be honest with yourself from this point on, you will wake up to all of the lies that you were made to think is true. Lies will no longer control your thinking. You *DO* have a choice. Choose to act according to the math fear cycle.

Or, you can choose to act in courage. Life has gifts to offer you. Discover your future without fear. Let's now remove your blindfold.

STEP 1: Be Aware of the Root Problem

Have you ever found yourself in a frustrating, never-ending cycle of confusion? I am sure that you wanted to end the frustration, somehow. But, found yourself in a similar situation months later. If we are not aware of the true source of the problem, we allow our disappointment to replay into self-destructive actions. Through ignorance, we repeat our trauma over and over again, expecting a different outcome. Our reoccurring, self-imposed, traumatic situations are disillusioned attempts to end our pain and fear. These actions prove to be unsuccessful, and we are left in a cycle of disappointment. Once we are aware of the true problem and choose to accept change, we move toward success.

Awareness is P-O-W-E-R-F-U-L. I wish I was more aware in my 20's. Perhaps, I could have avoided such crazy experiences. Though, I see that everything happens for a reason. I share this story, because, many times, we think that our happiness and success lies outside of our control. In fact, the truth is the exact opposite. We are the ones who create change in our lives by finding of the root cause of our frustration.

Break the Cycle by Awareness

Awareness is the first step toward breaking the cycle of frustration. In his book, Breaking the Pattern, Charles Stuart Platkin summarizes a 5-step process to break the cycle. Platkin explains that in order to break this debilitating cycle, we must follow this process:

A)   Understanding Our Patterns

B)   Understanding Our Past Failures

C)   Taking Responsibility For Our Actions

D)   Setting Goals

E)   Achievement

Although unconscious of this method, I used this process to be the first (and only) person in my family to graduate from a university. During the first day of my college career, my English professor, Johnie Scott, handed out an insightful study*. It revealed shocking statistics regarding the success of minority students pursuing higher education within the United States. The study found in 1993, that *1 in 31,000 African-American females would earn a degree in mathematics and the natural sciences.* This was especially a shock to me because I was enrolled as a math major! What were my chances of graduating? With this probability, I had a better chance of being hit by lightning twice than earning a math degree.

Needless to say, the odds were stacked against my favor. At that moment, I was determined to break my own cycle of self-sabotage. Years before, I was *Quincy* the Quitter and *Crystal* the Criticizer. During my tween and teen-age years, I had been an expert in procrastination and excuses. From seventh grade through high school, I felt inferior. I subconsciously questioned if I was as smart as other students. So, to save myself from future disappointment, I purposely waited until the last moment to do all of my homework. I thought that if I did not get a high score like the others, I had an excuse. My excuse was, "I did not have time to do the work."

In reality, I didn't have time to do the assignments. At 12-years-old, I was bussed into a school across town, and the travel time was horrendous. Plus, I was given adult responsibilities at too young of an age. I had to cook for my two younger sisters, prepare their laundry, and make sure their homework was completed. I didn't know that I also needed time for myself to develop.

Consequently, in high school, I did just enough to pass. My home high school was at least two years behind in its curriculum. And, the highest math class offered was Algebra I. If I attended my home school, I would have lacked the minimal entrance criteria for college.

Through my pre-college years, I used to *fake it* to get by. I faked it in English classes, and in History. I tried to fake it in Chemistry, but I earned a fail. But there is one class that I could never fake, which was my math class. Unlike the others, math always had one answer. However, the answer could be solved numerous ways. No matter how it was found, the question was always there to answer the problem. I gravitated to math, because out of all the subjects, it told me the truth. There was never a perception or "slant" involved in the answer. You either had it, or you did not.

Know the Mathaphobia® Infection Sources

Similarly, we can break the pattern of Mathaphobia® only if we know the root causes of it, and the pattern of how it is spread. With the brain's development, the brain tries to make sense of non-logical, dysfunctional situations. In doing so, the brain begins to malfunction from its intended operation. In turn, a child begins to internalize the dysfunctional situation as their fault. Sometimes, we are unaware that the root source of our own math confusion often comes from the ones closest to us. These false thoughts about math are implanted into our minds at early ages. The brain has been brainwashed by those already infected with math fears.

Richard Brodie, author of Virus of the Mind, explains a meme. A meme is an idea (true or false) that builds conclusions in the mind. In our minds, memes subtly influence our behavior. An example of a meme is a billboard advertisement with a girl on the beach having fun with friends, while holding a fast food-type hamburger in her hand. In this case, the meme is the idea that this burger equals fun. Memes are thoughts that are implanted into our brains.

Per Richard Brodie's research, memes are spread by influencing people's minds, and thus their behavior. Eventually, others are infected by the meme. Later, the meme turns into a mind virus. Like a virus, the meme penetrates the brain, copies itself, issues self-sabotaging instructions and spreads. Brodie continues to explain, "The mind viruses often fill our heads with self-sabotaging attitudes — memes — that hinder us from making the most of our lives," Brodie writes.

For another example, a meme can be the statement that your class member says, "This math problem is hard." This meme has now penetrated your brain. Next, this thought can be copied. As a result, you may think that other math problems may be difficult, as well. Thirdly, the meme mutates and issues a new set of instructions in your mind. Lastly, you may believe that you will not have enough time to solve the problem, because all of the problems are hard. Thus, you postpone your work. This meme spreads to your study group members. Now, this meme has turned into a mind virus, and the virus has diverted you from your original actions. The meme has you altering your actions due to false beliefs.

Mathaphobia® is the mind-virus relating the math and the actions surrounding harmful memes regarding math performance. This phobia gains a life independent of its origin and mutates to infect as many people as possible. The mathaphobia® mind virus can confuse and sabotage the brain and cause the brain to malfunction during the learning process. The spread of mathaphobia® is doomed to repeat the frustrating cycle of math illiteracy. Mathaphobia® is spread by parents, teachers and the school environment.

Root Source 1: Contagious Parents

A loud frustrating sigh explodes: "I can't do this! I don't understand it," yells Taylor, an 11-year-old boy, who wants to tackle his math homework. Yet, he feels overwhelmed whenever opening his textbook.

Barbara, Taylor's mother, sits down to help her son. She expects to understand his work. However, her eyes widen, while reading his homework. Barbara sees symbols and letters, which she vaguely remembers from more than 20 years ago. Barbara racks her brain, trying to recall simple math vocabulary words that she sees: *greatest common factor, least common multiple, undefined expressions*. Not only has Barbara forgotten the meanings, but she also realizes that she never learned some of the steps. In her own disappointment, she asks Taylor loudly, "Why haven't you been doing your homework? Hasn't your teacher been helping you?"

An even more frustrated expression overcomes Taylor's face. He feels like he can't gain help from anyone. With a seemingly bright idea, Barbara suggests that Taylor move to his *easier* History homework until she can ask the math teacher about the homework tomorrow. This appears like a smart idea. But with Barbara's hectic schedule, *tomorrow* turns into *next month*.

Barbara is unconsciously passing her own mathaphobia® on to her son.

The absence of a parental figure in the math education process creates perceptions of neglect. Too many parents do not spend time with their students. In turn, the student acquires and applies neglectful behavior to his or her own life. The student begins to neglect himself. He will spend little to no time on his own homework and educational self-development. Math homework is included in this category.

> *"The absence of a parental figure in the math education process creates perceptions of neglect."*

A parent's dysfunctional thinking is transferred to the student. The student thinks that if his own parent cannot spend time learning math with him, why should he spend time on his own self-development in math? This dysfunctional thinking directly correlates with *Quincy* the Quitter and *Samuel* the Struggler. The student begins to erroneously think he is not worth the time (or effort) to receive math directions from teachers or educators.

Root Source #2: Contagious Teachers

It is estimated that nine out of every 10 math teachers in the Los Angeles Unified School District (LAUSD) system, fail to hold a master's degree in mathematics. Furthermore, in the May 19, 2009 issue of *The Boston Globe* article titled, "Aspiring Teachers Fall Short on Math; Nearly 75 percent fail revamped section of state licensing test," it states three out of four or 75 percent of the new Massachusetts teachers looking to acquire their teaching certifications failed the math portion on the certification tests. Let's take a look at this extremity of locations, with respect of teachers from opposite ends of the United States. To me, the message is obvious. These instructors have not been formally trained in mathematical principals, concepts and notions of numbers and chance. Thus, it's possible such teachers are unable to effectively answer detailed math questions that students pose.

From my educational consulting work, I have observed teachers' responses. When they don't know the answer, some teachers ignore the students' questions. Or, in some dreadful cases, the teacher may give an incorrect answer. In either case, the student is left feeling confused and neglected. When more information is taught, the student is unable to digest it.

Math instructors, without the proper math education, are victims of mathaphobia® themselves. They fail to understand the main, intricate differences between math concepts. Their confusion is sensed by the entire class. I have witnessed countless mathaphobic® teachers, who begin to pause and hesitate in their speech, while attempting to teach. The teachers' board writing doesn't have math symbols and nor math details. Plus, the teachers have their back facing their classes for the majority of the lesson. Through these actions, the students acquire the emotions of terror from the teacher. In the end, mathaphobia® transfers from the teacher to the student.

The teachers' fears have a profound affect on students in college classes as well.

> *Lakshmi, a New York City Charter School Teacher elaborates on her thoughts: "Math and Science no longer get the same attention as English literacy in the curriculum. When I taught science in New York City, I was frustrated to see that the students did not have the math skills necessary to do the science problems. I had to teach math in the science period so that my students could do the work."
>
> *Martha, a Spring 2008 community college student shares her experience: "My high school Algebra teacher always made me feel stupid when I asked a math question. She literally ignored me when I asked another question. After a while, I stopped asking questions."

*Kent, a Fall 2007 graduate MBA student explains his experience: "It has been so long since I took a math class. When I returned back to for my graduate degree, I realized that I never learned the basics in college, or in high school. As a result, I struggled to keep up in my graduate statistics class. At that level, the instructors refuse to spend time with you to get it. – California MBA Graduate Student of Finance."

Root Source #3: Poor Math Curriculums and Mathaphobic® Peers

What happens when you give a piece of paper, and ask someone to read the following statement?

"A large pizza pie with 15 slices shared among 'p' students so that each student's share is 3 slices."

What would you think if the person could not read this in English? You would automatically think that something is wrong. That person would be considered illiterate. However, if you gave the equivalent statement in math:

$$"15/p = 3"$$

What would you think if the person could not read this? You may automatically think that the person hasn't seen math in years. So you may rationalize that this inability is okay. Would you probably think that he or she would have to "brush up?" Why is it okay to be illiterate in Math, and not in English? Innumeracy does not discriminate. Innumeracy spans every U.S. ethnicity, sex, age and household. Sadly, the requirement to take four years of high school mathematics is still a controversial subject in American K-12 curriculums! Math requirements are not as rigorous as English requirements. This controversy brings us to our next mathaphobia® culprit: Mathaphobic® School systems and peers.

In 2008 alone, China had over 300,000 college grads in math and science.

While in the U.S. there were less than 100,000. During the 2008 Beijing Olympics in China, *NBC Nightly News* Journalist Tom Brokaw aired a special report saying,*" Chinese math students believe destiny is in the numbers." In this broadcast, Brokaw reported that in China there is a mandate for junior high school children to take biology, chemistry, physics and math. Yet in the U.S., only 18 percent elect to take those subjects.

I remember my experiences in graduate school, and facing many other adversities in my life. I composed speeches based on this fact. For years I had the triple whammy: I am young, a minority woman, and a scientific female. Nonetheless, I discovered the fourth previously undetected adversity. I found myself to be the only American in a room of all foreign math students in STATE graduate school. Oftentimes, I was the only American in masters-level, math courses, and this experience was in the middle of Los Angeles, Calif.

And, when I participated in study groups, I felt like I was in another country: I would hear all Russian, Mandarin, Romanian, and even dialects of Indian. Nevertheless, the common language between all of us: Math. I enjoyed sharing a common bond of creative problem solving. In this learning process, I not only learned about math, but I also learned about different cultural characteristics. Most of all, I was surprised when I found out I was solving math slower compared to my foreign peers. They were taught math more efficiently and thoroughly, which allowed them to solve the problems quicker. On the other hand, I would struggle to understand basic concepts and theories. Then, I began to learn that my "advanced math" was inferior to the way in which they were taught.

By studying with foreigners, I learned that U.S. schools have the wrong emphasis in math education. U.S. schools emphasize "the right answer" and calculation. However, foreign students are taught to understand math theory.

Thus, I believe foreign students grasp theoretical concepts faster than their American counterparts. Plus, foreign math students are typically faster in testing and more sophisticated in innovation.

Proverbs 27:17 states, "Iron sharpeneth iron: so a man sharpeneth the countenance of his friend," (KJV).

> *"By studying with foreigners, I learned that U.S. schools have the wrong emphasis in math education."*

This sharpening was applicable to my math skills daily. By working with many cultures in the same environment, I received the opportunity to learn the foundations of math without fear. I was now in the company of others who *embraced* math. The opposite is also true. If I had stayed in my environment with others who dreaded math, I would not have been a rocket scientist. **When students are around peers who are afraid of math, they are conditioned to accept fear as if they, too, owned it!**

Step 2 - Surrender & Reprogram Yourself

$$E = mc^2$$

We can manipulate our minds to reject any type of fear. If we recognize that our thoughts have energy, then we can use this energy to restructure our brains for success. Our brains are powerful, and one scientist described this power. Albert Einstein is perhaps the most famous scientist of the 20th Century. One of his most well-known accomplishments is the mass–energy equivalence formula, $E = mc^2$. Although the formula is renowned, scores of people don't really comprehend that matter can be transformed into energy. Reversely, this energy can be transformed back into matter. Einstein discovered that energy (E) is directly related to the combination of both mass (m) and the speed of light (c). Ready for this, our thoughts have energy. And, it can restructure the mass in our brain.

In everyday matter, there are billions and trillions of atoms, which are the smallest particles in our universe. And, these atoms combine to form our brain, and its mass structure. The brain is by far the most complex organ we have in our bodies. It controls and categorizes all our experiences, emotions, thoughts, feelings and personality traits by using atoms enclosed in cells, which are called *neurons*.

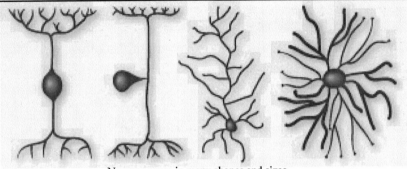

Neurons come in many shapes and sizes
(Source: Courtesy of http://learn.genetics.utah.edu/content/addiction/reward/)

Our brains contain over 100 billion neurons. They come in many shapes and sizes enabling them to perform specialized functions such as; storing memories or controlling our muscles. Each neuron carries atomic mass, and the transfer of the mass is seen in light waves. Transmitted thoughts and feelings can be represented by energy. So Einstein's theory applies to our thoughts as well. With distinct DNA structures, we each have a unique brain make-up. However, our brains have one aspect in common: we each contain explosive atoms in our head. Per Albert Einstein's mass–energy equivalence, I am convinced that our thoughts can be potentially explosive. The question is, "Which will it be: constructive or destructive energy?"

I am convinced we have a choice within our mental processes to choose to use our neuron atomic energy constructively or destructively toward our brains' development. In turn, our explosive thoughts can either contribute or destroy our personal success. You can use courage or fear to operate. If you are reading this book, I bet you are willing to change the way your brain fears math.

Dr. Wayne Dyer, an educational counselor and author of the book, Excuses Be Gone, justifies why we destroy our brains through explosive, negative thoughts toward our personal success. Dyer explains that most people habitually use excuses to prevent creative solutions in their lives. He states that re-programming our conscious mind is a simple process. First, we must be aware of what we are thinking. Dyer explains:

"The creative conscious mind can do almost anything you instruct it to do: you can change thoughts… Through discipline, effort, and continual practice, it can accomplish almost anything you focus your thoughts on…Your behaviors are supported by your thinking patterns; that is, your thoughts truly make or break your life."

He outlines the top 18 excuses. These "justifications" prevent frontal brain lobe neurons from forming. With the misuse of the brain energy, these excuses halt the brain from creative problem-solving. For the purposes of understanding the destructive mental attributes of mathaphobia®, I have adapted Dr. Wayne Dyer's excuses to math excuses. *Which math excuse do you use?*

Top 10 MATHAPHOBIA® Excuses

1. Learning math will be difficult.
2. Learning math will take a long time, because my scores place me in a beginning class.
3. I have others in my family who I must focus on.
4. Math is just not my strong point.
5. No one will help me learn math.
6. I was never good at math before, and I'm too scared to fail.

7. I am too old to learn what I should have learned when I was younger.

8. School policy places me in a remedial math class.

9. Learning math seems overwhelming.

10. I am too busy to learn math.

Once we have decided to reprogram our minds, our next task is critical. We must also be proactive and carry out actions based on our new thoughts! This is the only way our frontal lobes will reprogram and build new neuron pathways. We must channel the brain's energy into action.

Step 3 - Act with Health & Wealth

To act with health and wealth requires courage to do new things. I remember my time in church one Sunday afternoon. I felt horrible. I had $50 in my bank account, and I did not know where I would gain additional income. After resigning from difficult jobs, I believed that I placed myself back into poverty. Facing the natural struggles of life, I felt like I had lost my purpose. Although I possessed a large closet of clothes, shoes and jewelry, I still dressed frumpy. Despite possessing a relatively new car, I didn't regularly wash it. Even though I was paying rent for my Los Angeles apartment, I kept my living space in shambles. I was depressed, because I thought that I was poor again. Then, my pastor, Dr. Mark Brewer, stated a truth within his sermon. I held on to this "life-changing message" from that day forward. He explained that we often think that wealth only refers to money. But in fact, Dr. Brewer emphasized that affluence can be found in many forms. It can be viewed as an attractive appearance, charming personality, high intelligence, physical skill or a deep bank account. He continues to reveal that if we falsely think that our natural gifts are not valuable, then we are truly poor!

That moment, I recognized that I was wealthy! And, I was given exactly what I needed for my life's purpose. I realized that I was born with a natural ability to overcome every situation: once I was armed with the truth. My wealth was how I chose to communicate with people through my words and appearance. I also have the gift of empowering people with their innate abilities. However first, I had to notice my own wealth.

> *"Once we have decided to reprogram our minds, it is not enough to just think differently."*

The following days, I began to act with courage to claim my true wealth. I dusted off those nice pieces of clothing and polished my jewelry. Next, as a means of socializing, I found free events in my local area. Then, I became more thankful for having an apartment, and started regularly cleaning. Next, I invited people to my home for potluck dinners. I started to increase my base of positive friendships again. Despite only having $50 in my account, I felt rich.

Reprogramming continued, and my wealth consciousness transformed into wealth actions in business. I decided to focus my business on showcasing my public speaking skills. Previously, speaking bureaus rarely sent me out for speeches. With my new wealthy actions, I trusted that my courage to market my website would bring opportunities. I called up friends who managed different work establishments, and I asked to speak at their organizations. My efforts were successful. Later, I was booked for more paid speaking opportunities.

Once we have decided to reprogram our minds, it is not enough to just think differently. We must also carry out actions based on our courage to change. These courageous actions help our brains to support new neuron pathways to sustain a new way of operating.

CAPITALIZE ON YOUR OWN THINKING POWER

CHAPTER 5

CAPITALIZE ON YOUR OWN THINKING POWER

A t 11-years-old, I witnessed my mother bursting with contagious excitement. One Saturday afternoon, she returned home leaping up and down enthusiastically. Mommy had just taken a Myers-Briggs personality assessment in her Women's Studies Class. Beaming with joy, Mommy was overly thrilled, explaining how she was proud to be an "SJ!"

Naturally, I was clueless about her ecstatic emotions. But, as I sat talking with her, I began comprehending about how this unique psychological test helped Mommy understand how her brain operates differently from others. She felt empowered with this pivotal information. Now, Mommy was equipped with the right tools for expressing, identifying and building her self-worth. She discovered that her brain's uniqueness possessed a certain name. I, too, wanted to take that same test. But, I was too young! The school didn't administer the test to anyone under 18 because frontal lobes are still developing, and scores wouldn't be accurate.

However, this is what I found out: Your brain's personality can be measured.

In this chapter, you will gain the opportunity to take a free online type of Meyers-Briggs test; you will obtain insight on how your Meyers-Briggs personality shapes the way you operate in the world; and you will attain effective mathematic study tips based on your Myers-Briggs score.

The Myers-Briggs Type Indicator® (MBTI) assessment is a psychological questionnaire designed to measure psychological preferences in how people perceive the world, and make decisions. The original developers of the personality inventory were Katharine Cook Briggs and her daughter, Isabel Briggs Myers.

The test is constructed into four categories:

Attitudes: Extraversion (E) or Introversion (I)

Functions 1: Sensing (S) or iNtuition (N) and

Functions 2: Thinking (T) or Feeling (F)

Lifestyle: Judgment (J) or Perception (P)

Myers-Briggs Attitudes: Extraversion (E) / Introversion (I)

Briggs and Myers recognized that the brain operates in extroversion or introversion. Extroverted people tend to be energized by the external world of behavior, action, people, and things. People who prefer extraversion draw energy from action: they are more apt to act, than reflect, and then act again. If they are inactive, their motivation dies. They prefer to gain energy from interacting with the outside world.

On the other hand, the introverted personality is different. An introverted person acquires energy by being in a secure home environment. They rely on their internal world of ideas and reflection to restore energy. To rebuild their energy, these individuals reconstruct their energy through quiet time alone, away from activity. In short, extraverts are action-oriented, while introverts are thought-oriented.

Myers-Briggs Functions 1: Sensing (S) / iNtuition (N)

The first perceiving functions can be sensing or intuition. According to the Myers-Briggs typology model, each person uses one of these functions more dominantly than the other. This category describes how a person gains and understands new information. Individuals who prefer sensing are more likely to trust information that is tangible and concrete. They need to look for details and facts before they make a decision.

Individuals who prefer intuition view the data based on how it relates to a pattern or theory. They tend to trust their hunches. Those who prefer intuition tend to trust information that is more abstract or theoretical. Such individuals need to look for general patterns before they make a decision. They delight in future possibilities.

Myers-Briggs Functions 2: Thinking (T) / Feeling (F)

Thinking and feeling are the decision-making (judging) functions. The thinking and feeling functions are both used to make rational decisions, based on the data received from their sensing or intuition function. Those who prefer the thinking function tend to make decisions with logical and consistent rules. They do not place much value on personal feelings in this process. Those who prefer the feeling function tend to make decisions by empathizing with the situation, and people involved in the situation. And, they value harmony from the decision that is made.

Myers-Briggs Lifestyle: Judgment (J) / Perception (P)

Myers-Briggs identifies that people also have a preference for judging *or* perceiving. Judging-types typically like to have a time schedule and situations defined. However, those who have the perceiving-thinking type like to operate with a flexible time schedule, and prefer to keep decisions open.

The four functions operate along with the attitudes (extraversion and introversion), and the lifestyle (judgment or perception) of a person. Each function is used in either an extraverted or introverted way. A person whose dominant function is an extraverted feeling, for example, uses intuition very differently from someone whose dominant function is introverted feeling.

> *"Each personality shows how we diversely use our brains to creatively problem-solve."*

For the comprehensive Myers-Briggs test, you can go to a college testing center for the report. Or, you can acquire free online reports. Although, these free reports may be basic in its analysis. You can obtain a free, comparable Myers-Briggs personality assessment from many online websites. One free website is *http://www.humanmetrics.com/cgi-win/jtypes2.asp*. There are more websites that can offer your personality type:

www.mypersonality.info

www.personalitypage.com

www.keirsey.com

www.typelogic.com

www. webspace.webring.com/people/cl/lifexplore/

Overall, there are 16 different personality results from the Myers-Briggs personality test. These categories are grouped into four assorted categories.

No category is better than the other. They are equal within brain operation and intelligence. Each personality shows how we diversely use our brains to creatively problem-solve.

Now, I will give you a quick summary about the basic meanings between the names in each category:

I think differently than you. And you have different experiences than I. However, we all have a common human brain. We have unique, yet similar, thinking skills. We must know how to use our personalities to benefit and feed our brains. It is our job to optimize our brain functions. In school, teachers do not offer us these tools. I will shed light on the process to master the mind and body connection, while studying math. In this section, we will see how the brain's personality best learns math.

Different Myers-Briggs Personality Characteristics

= Different Math Study Techniques

SJs Temperament – *"The Protectors"*

If you have an ISTJ, ESTJ, ISFJ, and ESFJ, then you have a
***SJ Temperament*, also known as "The Protectors."**

SJs are observant, stable and motivated by a need to be placed in consistent situations. They believe in a steady routine, and are excellent at completing tasks. SJs are polite and straightforward, respectful and more formal than others. They feel as if people should behave well-mannered at all times! SJs are thorough in their plans, and enjoy when everything is completed accordingly. If a need is justified, SJs are quick to provide a solution when problems arise. They speak of facts and precise details, rather than generalizations. SJs ensure no more and no less credit is given than what is due. SJs steer away from impulsive actions, and aim toward gaining concrete facts before taking any moves. SJs are reliable, hard-working, solution-oriented people. They prefer to verify that a tradition is followed. SJs make up the greatest part of the management of business organizations. Their attention to detail lends well to the structure of larger corporations.

SJs Typically:

- Embrace a strong sense of responsibility and duty.

- Practical, traditional, and organized.

- Not interested in theory or abstraction unless they see the practical application.

- Able to see clear visions of the way things should be. (If they are Extroverted.)

- Loyal and hard-working.

- Exceptionally-capable in organizing activities.

- Dependable and have follow-through tactics.

- Stable and practical: they value security and traditions.

- Hold a rich, inner-world of observations about people. (If they are Introverted)

Study Tips for "The Protectors"

If you are an SJ, there is a study process tailored for you:

1. **Write out the steps in each math problem and then, number each problem.** SJ individuals thrive when they have step-by-step directions to carry out. Without steps, SJs feel like they are looking for a needle in a haystack. As a result, a good study tip is for SJs to write out the steps to a math solution. If asked to solve for x in the following problem:

$$4x^2-x = 0$$

You can write out the steps in the following manner:

Step 1: Factor $\quad\quad\quad\quad\quad\quad\quad\quad$ **x(4x-1) = 0**

Step 2: Set each quantity equal to zero $\quad$ **x = 0 and 4x-1 = 0**

Step 3: Solve for x $\quad\quad\quad\quad\quad\quad\quad$ **x=0 and x = 1/4**

For example: Step 1 would be to factor. Step 2 would be to set each quantity equal to zero. Step 3 is Solve for x. When you have steps, it clears up the confusion, allowing you to focus strictly on the solution.

2. **Do additional homework problems, if available.** You have a great sense of work ethic. Once you repeat the processes, it permanently lodges into your brain. It is a good idea to ask the teacher for additional math problems, if you need more information about a concept. If you do at least three problems from each section, you will understand the concept quickly.

3. **Go to the teacher's office hours to ask questions.** It is likely that you will need to understand certain concepts. Or, you will need more problems to practice. Use this opportunity to see the teacher, and ask for specific steps. As a teacher, I find myself giving more information to the ones who visit me during my office hours. When you take the time to go directly to the teacher, you also have the opportunity to gain more insight than just the lecture alone. You will be able to pick the teacher's brain. Quite frequently, (shhh ... don't say that I told you this information) teachers usually give major hints toward future test questions to the students who visit the day before the test. Most teachers sincerely want to see their students do well on their exams.

4. **When asking the teacher about the problem, request the step-by-step method to the solution.** When you receive the step-by-step solution from the teacher, the teacher has the opportunity to explain it to you in more detail. The higher level of detail that you have regarding the math solutions: the more likely you will do well on the exam. Automatically, this refers to step 1, listed above.

5. Learn the math sections in chronological order. How do you feel if someone tells a story without giving you the full detail? More than likely, you will feel quite frustrated. The same is true for learning math. If you skip sections, you will always feel like you have missed something important. The trick is to read every section that you cover in your class. You do not have to workout problems in each section. However, if you know the concept in each section, you will have a more complete feeling that you can comprehend the concept.

> *"When you have steps, it clears up the confusion, allowing you to focus strictly on the solution."*

6. Keep neat and clean lecture notes, and date each lecture. If you are an SJ, you enjoy order. When you have a neat house, you more than likely, can think better at home. The same is true for your notebook. If you keep lecture notes dated, organized, and in place, you will remember what you wrote down on your paper. If your notes are chronological, you will remember detail with ease. Let's say that you do not date each lecture, your mind will be fixated on the date that you saw the math, rather than the concept.

7. Participate in a study group with no more than 3 people. If you are an SJ, you like structure to your time. If you study with more than three people at a time, then too much time will be spent trying to communicate. Rather than spending time comprehending the math, limit the study group to no more than three people. Then, you will be able to keep your studying schedule on track. And, you will all feel a sense of accomplishment afterwards.

NT Temperament – *"The Intellectuals"*

If you are an ENTP, INTJ, ENTJ, or an INTP, then you have an *NT Temperament,* also known as "The Intellectuals."

NTs enjoy pondering about ideas. They can also be described as being analytical, intellectual, inquisitive, independent, and complex. NTs thrive on finding solutions that no one has yet discovered. Such intellectuals are known for inventing algorithms, and devising strategic solutions. Most scientists and engineers are NTs. This introspective, logical group is known for being on a constant quest for knowledge. NTs are focused on function: practical and unsentimental in their approach to problem-solving. In work assignments, their main focus is to "complete the mission at any cost." The actions of an NT are precise and follow a systematic logic: abstract, theoretical and technically-adept. NTs are naturally-inventive. They typically have hobbies, which include studying, self-improvement, and learning about new things or languages. This logical group typically makes excellent CEOs and organization leaders.

NTs include the ENTP, INTJ, ENTJ, and INTP. NTs Typically:

• Independent, original, analytical, and determined.

• Able to turn theories into solid plans of action.

• Appreciative of knowledge, competence, and structure.

• Driven to derive meaning from their visions.

• Long-range thinkers.

• In possession of very high standards for their performance, and the performance of others.

• Natural leaders, but will follow if they trust existing leaders.

• Excited about new ideas and projects, but may neglect the more routine aspects of life.

• Able to understand concepts and apply logic to find solutions.

Study Tips for "The Intellectuals"

If you are a NT, there is a study process tailored for you:

1. **Talk through the math steps in each math exercise to see the approach.** As a NT you are strategic. This means you need to know the strategy within math problems as well. When you speak math aloud, you use two portions of your brain. Therefore, your brain digests the math concepts with more strategic rigor. While speaking aloud, your brain equates speaking with leadership. So, when you speak math aloud, you have a higher appreciation over your math comprehension. Unlike the SJ where you must write the math steps down, you need to speak out the steps. Thus, you obtain a mental

organization as you view the math. This mental organization allows you to categorize the concepts in your head.

2. **Differentiate between the various strategies to tackle the same math concept.** As an NT, you value long-range thinking. In this spirit, when you are given a math problem, ask the teacher about how and when you would use a strategy. Here is an example:

$$4x^2 - x = 0$$

You may not know if you have to use the quadratic formula first. Or, factor out the greatest common factor. In this case, you would ask about the best strategy to tackle this problem. As a result, the teacher will explain the Rule of Thumb for factoring this problem. Always factor out the greatest common factor first. Do this step, if possible before using the quadratic formula.

3. **For stronger memory tactics, relate each math concept to other real-life experiences that are familiar.** My personality type is an ENTJ. I struggled to remember certain math concepts. But, one real-life encounter helped me learn how the concept of direct proportional relationships works. I learned how to understand proportional relationships when I related my personal experience with the boiling point of water and salt saturation. The following story demonstrates my explanation:

I was 8 years-old, and my older sister loved playing practical jokes on me. She said that if I wanted my Malt-O-Meal™ water to boil faster, I had to add tons of salt to the water. Of course, she lied. I eventually waited two hours for two cups of water to boil!

I placed too much salt into the water. Needless to say, the Malt-O-Meal™ was disgusting. But this real-life concept showed me a direct proportional relationship: the more salt I added: the more time was required to boil the water. This story always helps me to understand direct proportional relationships. More salt resulted in a higher boiling point.

4. **Organize study groups with 2-5 people, with 5 being the maximum number of people.** If you are a NT, you are more than likely great at managing projects and groups. Most corporations enjoy hiring a NT personality, because NTs are known for coordinating people for tasks. This personal quality is the same for math study groups. When you study in groups of 2-5 people, you are able to smoothly intermingle between conversations about certain topics. However, NTs are great at starting a math problem, but have a tough time with completion. You can pair with an SJ who is able to help you finish the problem.

5. **Ask questions during the class lectures.** Any and every question that you ask is important. Your education is valuable. Never be fooled. Your questions are worth answers. As a NT, you typically have true, thought-provoking, strategy-minded questions. This is how your brain works. And, you probably represent five-to-six other class members who have the same concerns. Be courageous and ask questions in mid-lecture. Always be respectful. If the teacher gets offended, tough luck. It is the teacher's job to answer your questions. You are the customer in the education business system.

6. **Ask the teacher for the strategic approach in each exercise.** As a NT, the main question that you should ask is, "Why?" Ask this

question more than "Who," "What," "When," or "Where?" Your brain neurons trigger faster when a "Why" question is answered. Then, ask the other questions once you understand, *"Why on Earth?"* you are doing these math steps.

NF Temperament — *"The Visionary"*

If you have an ENFP, INFJ, ENFJ, or the INFP personality, then you have an *NF Temperament*, also known as "The Visionary."

NFs tend to approach life and work in a warm and enthusiastic manner. The optimistic personalities like to focus on ideas and possibilities, particularly, "possibilities for people." They are often found in careers that require communication skills, and focus on the abstract, along with an understanding of others. They dislike careers, impersonal analysis, technical approaches, and factual data: NFs prefer the human mind and its emotions. NFs are often found in the arts, clergy, counseling, and psychology, writing, education, research, and health care. The NFs are intuitive with people, and are highly-idealistic. They are subjective, compassionate "feeler" people that desire to contribute goodness and meaning to the lives of others. They are effective nurturers and encouragers. NFs despise conflict, and will do everything they can to create harmony. They are imaginative, creatively-inclined, and passionate about their causes.

NFs include the ENFP, INFJ, ENFJ, and the INFP. NFs Typically are:

- Externally-focused, with real concern for how others think.

- Able to see everything from the human angle, and dislike impersonal analysis.

- Interested in serving others.

- Popular and sensitive, with outstanding people skills. (If they are Extroverted.)

- Enthusiastic, idealistic, and creative. (If they are Perceiving.)

- Able to do almost anything that interests them.

<u>**Study Tips for "The Visionary"**</u>

If you are an NF, there is a study process tailored for you:

1. **Write out a short story and relate it to each problem.** If you are an NF, you will remember things that happened in life based on relationships. For example, the order of operations is Parenthesis first, then exponents, multiplication, division, addition and subtraction. If you were to remember the first letter of each operation as PEMDAS, then you could create a story to remember the order. With this approach, you will have a clever way to make math into a personal story that you will not forget.

<u>P</u>lease <u>E</u>xcuse <u>M</u>y <u>D</u>ear <u>A</u>unt <u>S</u>ally.

2. Meet with study groups comprised of 2-3 people; 3 being the maximum (including you). As an NF, you thrive on people interaction. If you are in a group with more than three people, you will feel deprived of the personal interaction within studying. When you have an opportunity to combine socializing with studying, you will equate math as a pleasurable experience. Your mindset toward studying changes, because you will know that you are not alone in this process.

> *Spending time with people who know how to create various math solutions, enables you to "pick the brains" of those stronger in math.*

3. Ask the teacher questions before the class, and during his/her office hours. Spend time with people who know how to create various math solutions. As an NF, you thrive on understanding people. If you meet with the instructor before class, you have the opportunity to connect with the teacher. During this time, you can ask the teacher to cover certain topics, as well as answer your questions during lectures. By frequently visiting your teacher, it also guarantees your teacher remembers you. Spending time with people who know how to create various math solutions, enables you to "pick the brains" of those stronger in math. Hence, you remember other thinking processes since it will be associated with personal interactions.

4. Ask the teacher for the difficulty levels of each math exercise. As NFs, your brain thrives off of strategies. You are not solely focused on step-by-step approaches. Rather, you are interested in the answer to "Why" and "When?" Use your thoughts for your advantage. Create a difficulty level for each type of problem in a math section. A "1" represents the easiest, and a "5" represents the most time and strategy a math solution will require. If you need help determining this, ask your teacher which math problem is simpler. For example, let's say that you are studying exponents. If you were asked to simplify exponential expressions, you can write the level of difficulty next to each problem:

- **(1 = Most Straight forward)** $x^6 x^5$

- **(2 = Less Straight forward)** $(x) (x^7 x)^6$

- **(3 = Medium)** $(x) (x^7 +1)$

- **(4 = Some clever thinking involved)** $(x^7 +1) (x^7 +1)$

- **(5 = Most clever thinking involved)** $(x^7 +1)^2 (x^7 +1)$

5. **Serve as a resource for other students (notes, phone contacts, etc.)** If you are an extraverted person, an option exists for you since you are a friendly person. If you were to obtain each class member's phone number and email address, you could be known as the "go-to" guy/gal for class information. You could keep notes and assignments handy from lectures. This approach does require commitment. However, you could give this information in exchange for the opportunity to be involved in any study group held.

SP Temperament – The *"Creators"*

If you have an ESFP, ESTP, ISTP, or the ISFP, temperament, then you have *SP Temperament*, also known as "The Creators."

SPs are observant, experimental and primarily driven by sensation. They are the artists and performers of society! SPs typically tend to "live in the moment," get along easily with people, are easy going and carefree. SPs dislike rules and guidelines, and gravitate toward hands-on approaches. They are impulsive and seek the next experience or adventure. SPs often seize opportunities that may prove to be thrilling, and pleasing. They are flexible, aesthetically-aware, "here-and-now" people that tend to go where their senses lead. SPs don't spend much time planning or philosophizing, preferring to take spontaneous action. SPs are laid-back, open-minded, and always need to experience the beauty of life. They have a tendency toward athletics and anything that involves creating or crafting.

SP's include the ESFP, ESTP, ISTP, and the ISFP. The SP's typically:

- Possess a people-oriented and fun-loving personality.

- Make things more amusing for others by their enjoyment.

- Live for the moment: they love new experiences.

- Dislike theory and impersonal analysis.

- The center of attention in social situations. (If they are Extroverted.)

- Impatient with listening to long explanations. (If they are Extroverted.)

- Have a well-developed common sense and practical ability.

Study Tips for "The Creators"

If you are a SP, there is a study process tailored for you:

1. **Create index cards for each vocabulary word, exercise and topic discussed in class.** As a SP, you follow the beat of your own drummer. This means that you follow your own time schedule. Your flexibility will allow you to study anywhere at any time. You can create flash cards for your math vocabulary words. One side can be the term, and the other side can be the definition with a sample exercise. Set a goal to relearn 5 to 10 new items per day. Set a goal to write down 20 new flash cards a day. Pull these cards out anywhere, and study at your own pace.

2. **Relearn the flash cards for 10-15 minutes apiece.** When you prioritize time for each concept, the approach will allow your brain to digest the math concept. The concept is seen at least three times: once from the lecture; next from the flash card creation; and last from the self-quiz. Three times is the lucky charm.

3. **Ask questions before, during, or after class.** As a SP, you are flexible with time. So, you can ask questions at any time. Though, you have to promise yourself to actually ask the questions when they come to your mind.

4. **Study math with instrumental music.** If you are a SP, you probably enjoy music more than the average person. When you listen to music, you profit from unlimited possibilities. Now, you can transfer that feeling into studying math as well. Study math, while listening to instrumental versions of jazz, classical, pop, or any of your favorite music genres. Math and music stimulate the same frontal brain lobes for SPs. Instrumental music allows your brain to focus on creation instead of fear.

I had an interesting experience with one of my tutoring students, while assisting in her transition from junior college to a four-year university.

Ana

As a returning older student, Ana struggled to learn math because she didn't know how her brain worked. This former professional at an international music awards company, was now attending a local community college. Ana was hoping to transfer to the nearby four-year university, and earn her Bachelor of Arts degree. Aspiring to be a role model for her adult daughter also attending junior college, Ana was a student in my Beginning Algebra class. Quiet in class, Ana would frequently become lost within the lecture. However, she visited me during office hours, and passed with a "B." The next semester, she contacted me for help. She was failing Intermediate Algebra, and said she had a poor teacher. Ana also said that the teacher went too fast. She said she needed a better teacher or tutor so she could understand the issues. I agreed to help Ana.

As I worked with her, I noticed that she was really determined to do well. However, when I showed her math concepts, she couldn't focus. She completely tuned out from the lessons. Ana was busy worrying about concepts from the previous sessions. Similarly, she didn't hear the math lectures, even though she was present in class. Ana couldn't tell me about the lecture or what was happening during tutoring sessions. Oftentimes, Ana thought the teacher never covered certain subjects. Ironically, I discovered Ana's teacher had lectured on the very topics that left her confused. Instantly breaking into tears, Ana often felt overwhelmed with math problems, especially when she saw fractions. Whenever she did feel confident about particular math concepts, she rushed to solve the answers quickly. Frequently during this rush, she made logical errors in her mathematical steps.

After working with Ana, I discovered her true dilemmas.

1) A learning disability.

2) A traumatic math history.

3) A need to be perfect.

With Ana's help, a plan was designed for her overall academic collegiate success. First off, Ana registered for testing resources through the campus Special Services Center. Next, we found a classmate to take notes for Ana, while she watched and listened during the lectures. (Like many special needs students, Ana is unable to multi-task at this level.) Then, we set realistic, weekly time frames for Ana to finish tests, and study for classes.

More importantly, I discovered the life-changing event that created Ana's initial mathaphobia®, paralyzing her everyday thoughts. One afternoon, she tearfully told of her horrific, sixth-grade math teacher. Whenever a student miscalculated a fraction, the teacher proceeded to publicly scold and put down the student. Since then, each time Ana sees a problem involving fractions, she completely freezes. With this finding, we revisited the theory behind fractions. Ana became comfortable with fractions and could tackle math problems.

Lastly, I found during the sessions and within her math class, Ana feared her mistakes would make the teacher no longer like her. (This behavior is similar to her sixth-grade trauma). Therefore, she would become distracted, focusing only on the topics she did not understand during the lectures. This made her miss important information. To combat this issue, I identified her Myers-Briggs personality type and offered her methods to study with others. This information also helped her to communicate with the instructor. Consequently, Ana felt more confident with her math knowledge, and focused better during lectures.

Upon activating the full strategic plan, Ana was able to pass her developmental math course, along with successfully completing Business Calculus! She is well on her way to earning an Economics degree from a four-year university.

As you can see, the Myers-Briggs Type Indicator® shows how your brain processes information. When we understand how our brain works, we can capitalize on specific methods that increase our math comprehension. Using the four distinct techniques, we are able to effectively communicate mathematics to the outside world, as well as become effective math practitioners and test takers. With this information that I share with you, I am just as excited as my mother was when she returned home from her Women's Studies Class! But, knowing these study tips alone will not help you ace math. You must choose to eliminate the mathaphobia® that you have. Within the next four chapters, you will know the origination of your mathaphobia®. You will also learn how to reprogram your brain, and take action steps to remove your fear. Are you ready for the next step in your journey?

Turn Quincy the Quitter into David the Determined

CHAPTER 6

TURN QUINCY THE QUITTER INTO DAVID THE DETERMINED

Two roads diverged in a wood, and I,

Took the one less traveled by,

And that has made all the difference.

- An excerpt of the poem, "The Road Not Taken" by Robert Frost

Question: Tell me. What do Dr. Martin Luther King, Jr., Abraham Lincoln, and Mother Teresa, all have in common?

Answer: They chose the path less traveled although the odds were against their favor.

I n life, you have no guarantees. As mentioned before, I had a 1 out of 31,000 chance of earning a Ph.D. in Mathematics or the sciences! I could have easily chosen to give up, and change my major. Instead, I decided to move along the road to success. I chose the road *rarely* traveled. If I had stopped trying to understand math when I first failed Algebra in 8th grade, would I have earned a master's degree in mathematics? Do you think I would eventually have become a rocket scientist if I had abandoned math when I failed Geometry in 9th grade? Would I have appeared on TV as a Math and Science expert if I had quit learning math after receiving a "D" in my Calculus class? Of course, there were many stumbling blocks along my journey. But, I chose to be determined, stepping on those blocks like stones: even though it was easier to quit.

In this chapter, if you are *Quincy*, you will learn to turn away from a "give-up" attitude by understanding the root cause of your crippling thoughts; re-programming your brain to expect a triumphant end; and by equipping you with the tools to face future math classes. You will also learn to become David the Determined. To be determined again, we will follow three steps:

Step 1: Know *Quincy's* Root Issue

We all have problems in life, and we must find the root source of our dilemmas. *Quincy* is no different. He needs to understand why he automatically kills his own success. *Quincy* believes that he will fail before he tries. An amazing person, but in order to get to the root of *Quincy's* issues, we must first consider these basic questions. Take a moment to close your eyes (if possible), sit back, and relax. Travel back in your memory and recall the following:

At what age did you experience your first major disappointment? It can be any event, math or unrelated to math.

What happened when you tried to gain help in that situation? What was the outcome? Explain three, self-defeating thoughts you created at that moment that impacted your choices.

How did you feel at that very moment? Explain those feelings on paper or say it aloud.

When did those self-defeating feelings reoccur? Did those feelings occur in a math class, during tutoring or at home?

When you thought about failing or not understanding math, how did you cope with those feelings?

Pinpoint a distinct, subsequent incident when you decided to give up before trying.

Remember that moment in time. Say aloud:

> I cancel that moment in time. I forgive myself for choos-
> ing to accept fear in my past. I will no longer choose to
> be ruled by the fear that tries to creep into my mind. Fear,
> you are no longer welcomed into my life!

Now, decide to change the outcome. Return to that moment in time, and pretend that you have nothing to fear. What would you say to that person? Say it.

Pretend that you chose to stay determined to understand math, and get help.
Describe how different your actions would have been.

Explain three, encouraging thoughts that you could have possessed during the
moment that positively impacted your choices.

Open your eyes, and come back to this moment.

When *Quincy* views math, he sees himself as a failure. Secretly, he is overwhelmed. He fears making mistakes. It's easier for him to remain silent in
math class, rather than tackle all of the self-defeating thoughts running through
his head.

> *"Fearful that he'll be ignored and neglected when reaching out for help, Quincy simply doesn't ask."*

Quincy doesn't want anyone to know he has mathaphobia® issues.

Fearful that he'll be ignored and neglected when reaching out for help, *Quincy* simply doesn't ask. As a result, *Quincy* has mastered avoidance behavior. There is good news: we can change *Quincy*'s self-defeating thoughts and actions. But, in order to do so, *Quincy* must ask himself some motivational questions:

Questions for *Quincy*

What activities have you mastered (i.e. sports, auto mechanics, cooking, surfing, sewing, etc.)?

What three skills did you learn to help master that activity?

Who coached you with these accomplishments?

Name three things that the coach did to help you believe in yourself?

What amount of time do you allocate to mastering your skill (weekly, daily, and monthly)?

Become David the Determined

In one sentence, David turns his fear into focus.

David and *Quincy* have the same mind, but David reprograms his mind to accomplish his goals. David is a champion. Once David finds an obstacle, he is determined to overcome it. David blocks insults and harassment, and replaces both with empowering actions that help him triumphantly succeed. He is a thinker and has a curious, inquiring mind. David has excellent reasoning powers, and is not stubborn. David works to maintain his concentration, discipline, and focus. And he knows what he can do, verses what he cannot. He pays attention to practical matters, following through on his promises

> *"David avoids bouts of confusion, because he finds a mentor who helps guide him. "*

David is sensitive, but more vulnerable when someone implies that he cannot complete a goal. He maintains a calm disposition when questioned. If David makes a mistake, he forgives himself, which in turn fuels him to complete his purpose. David has ambition, organization and practicality. He is concerned with achieving results, rather than simply letting life's factors overrule him. To avoid letting his emotions take over, David focuses on long-term plans. He envisions his own success before it happens.

David avoids bouts of confusion, because he finds a mentor who helps guide him. David realizes overly-high expectations can be unrealistic, and can prevent him from accomplishing goals. He knows with the right supporters, his abilities can reach their full potential. David materializes his dreams based on his self-created, step-by-step, action plans. An independent thinker, David is a natural leader that's goal-driven.

Step 2: Divorce Avoidance & Re-Marry Help

Remember, when *Quincy* views math, he sees himself as a failure. Therefore, *Quincy* has mastered avoidance behavior. However, David is a champion, and he knows how to gain the help to succeed. As a result, David transforms *Quincy*'s thoughts, so *Quincy* can become a champion, too. With a mind reprogramming technique, David blocks *Quincy*'s self-insults and self-harassment, replacing each with bold and inspiring thoughts that are said aloud.

- *Quincy* thinks, "I am never going to understand this math."
- David Reprograms, "I will understand parts of math, and eventually most math topics."

- *Quincy* thinks, "I am not that smart."
- David Reprograms, "When I am determined, I can understand anything."

- *Quincy* thinks, "No one will want to help me understand it."
- David Reprograms, "There is always someone who is willing to help me. I must find them."

- *Quincy* thinks, "I fail at things in my life."
- David Reprograms, "With my courage, I will start, commit, and succeed in special areas of math."

- *Quincy* thinks, "I will never be successful, so why should I try?"
- David Reprograms, "When I am determined, there is nothing that can stop me."

David's Daily Affirmations:

I will be acknowledged.

Every challenge I face: I will succeed.

Every day, I am stronger in math.

Every moment, I am more determined to succeed than before.

I strive to do my best in math, and in all math concepts that I attempt.

I am determined to finish my math class with the best score that I can earn.

I am destined for greatness.

If I start, I never fail.

I create my opportunities.

Step 3: Take TRIZ® Actions Steps

"TRIZ®" is the (Russian) acronym for the "Theory of Inventive Problem Solving." TRIZ® is an international science of creativity that relies on the study of patterns of problems and solutions, and not on the spontaneous creativity of individuals. TRIZ® is a problem-solving method based on logic and data, not intuition. G.S. Altshuller and his colleagues in the former U.S.S.R. developed the method between 1946 and 1985. More than three million patents have been analyzed to discover the patterns that predict breakthrough solutions to problems.

Within the TRIZ® methods, there are three main principles; 1) Problems and solutions are repeated across industries and sciences; 2) The same pattern of innovation is repeated across industries and sciences; and 3) Creative innovations use scientific effects that can be from a completely different field.

There are 40 principals in TRIZ used to find solutions to difficult problems.

For example, one of its principles uniquely finds solutions to one problem by extracting examples from a similar problem in another field. Consider this dilemma. Let us say that a farmer wants to grow a tree and produce fruit. But, the tree was planted in a calm environment. As a result, the tree was not strong enough to bare fruit.

Using TRIZ® methods, we could take a solution from physical therapy to solve the issue. When people age, the spine requires strengthening to effectively carry its body. One approach to strengthen the spine is to provide low levels of weight resistance. In this way, the muscles formed through light-weight lifting serves to strengthen the spine.

> *"Again, it is not sufficient to change the way that Quincy thinks, but he must also change the way he operates in order for new brain neurons to form."*

If we apply the solution from physiology to farming, then we can gain true results. When new trees are planted in a slight wind breeze, the trees will be strengthened by the resistance of the wind. As a result, future trees bore fruit. The main point of this story is that, IF we use solutions from other areas in our lives, then we can solve our issues within math too. Here are three steps that we can use.

Use TRIZ® to Cure *Quincy*

After studying this TRIZ method, I realized that we can use its principles to solve *Quincy*'s issues. At the same time, *Quincy* can transform into David. Again, it is not sufficient to change the way that *Quincy* thinks, but he must also change the way he operates in order for new brain neurons to form. Below are four steps that can transform *Quincy* into David.

Step A: Re-Apply Actions that Work

- **Recall all the tasks where you excelled before, and choose to apply that same effort in math courses.**

Think back. Were you an excellent athlete? Or, were you an excellent musician? Oftentimes, the same steps that we took to master other tasks can be used to master Math.

Consider the real-life story of Jose.

The Story of Jose C.

Jose C., a returning, mid-20's student in my college Algebra class, possessed a tall, athletic frame existing in polar opposition of his shy personality. He took his first test on fractions, and failed miserably. I gave students the opportunity to take their graded tests home and work with a tutor for their corrections. Jose C. did well on the assignment. However, when I issued the next test, Jose C. also failed that one. I did not understand this phenomenon.

Shortly after the second test, I furnished my entire class information about mathaphobia®. I defined each of the mathaphobia® characters, and asked the students to take the self-quiz. Jose took the test and discovered he was *Quincy* the Quitter. As I described *Quincy*'s characteristics to the class, Jose C. recognized that he was *Quincy*. I then asked class members to share their story, and Jose C. shared his past.

Jose C. revealed that he was an award-winning football athlete in school. He won several championships. I asked about this training method, and he explained how he met daily with his coach at 4 p.m. His coach showed him how to catch the ball, run, and how to anticipate the opponent's next moves. Then, Jose C. says he would go home and practice those moves. He verified that each week, time was set aside for his home practice. While at home, Jose C. had family members pass the football with him for practice. And before the game, he envisioned himself successfully catching the ball for touchdowns.

After listening to Jose C.'s story, I realized the exact steps required for mastering football were the identical steps needed for Jose C. to master math. Jose C. explained that everyday at 4 p.m., he trained with his coach. Similarly, I asked Jose C. if he could train daily with a math tutor on campus in the afternoons. Most college campuses have complementary math tutoring available for students. I explained to Jose C. how his math tutor could serve as his "math coach." The tutor could show Jose how to view problems, use the Order of Operations, and show the fastest way to solve problems. Jose C. agreed to meet at 3 p.m.

Next, Jose C. verified that each week, he could set aside time for home practice. Many students don't know that for every one hour of college class, they're expected to study an additional three hours outside of class.

This is true for any college class that college students take. Since Jose C. scheduled an hour each day with the tutor, he found an additional hour each day, (Monday through Thursday), to study math on his own. He gave himself the weekend off.

> *"Many students don't know that for every one hour of college class, they're expected to study an additional three hours outside of class."*

Similarly, Jose C. began meeting classmates to practice math problems. They would give each other questions, and then take turns finding the answers. Oftentimes, they would combine their knowledge to be able to complete each problem. Jose C. formed this group and it came to serve as his support group, and he felt that he could do the math with their encouragement.

Lastly, Jose C. learned that he had to expect personal success. His performance could reflect his effort. Before every test, Jose C. closed his eyes and imagined himself receiving the test paper, and tackling math problems. Similar to how he envisioned himself catching footballs and running for touchdowns! Jose C. imagined himself tackling the math problems, and using all the operations he learned during math tutoring.

After Jose C. discovered his game-plan for math, he went into action. He took the third test, and he earned a high "B" score. On the fourth exam, Jose C. reached closer to the end zone, scoring an "A." When he took the final, Jose C. was the first to finish, answering nearly every question correct! Touchdown! At the end of the class, he thanked me for showing him how not to be a quitter, but a champion in Math.

Step B: Take "Mr. Help's Class"

- **Select instructors who are known for helping students.**

Most likely, if you relate to *Quincy*, you've been afraid to approach math teachers and share your experiences of feeling lost. The fear increases if you have a teacher known for being hard, and expecting students to learn math on their own. Or, even worse, the teacher may speak with such a heavy accent that you're unable to understand him or her. These situations suck! To avoid such circumstances, you can enroll in Mr. Help's (or Ms. Help's) class.

Mr. Help is on every campus. This female or male teacher actively "helps" students learn. Let me state this again: EVERY CAMPUS has one of these teachers. They make themselves available for questions in their office. They offer additional resources to assist you in numerous subjects. Ms. Help suggests tutors for you. Typically, they are extremely patient, and speak slowly while teaching.

To find Mr. Help's class, ask around your campus. Most people are willing to share information about teachers who they've enjoyed. Also, be weary of websites with teacher ratings. Sometimes, the only students that go on these sites are the students who did not like a particular teacher. And these students log into these websites under several aliases for the purpose of writing bad reviews. Hence, the website ratings may not be factual reflections. Asking near your campus' math departments should provide more accurate descriptions. If all else fails, go to the math department office and ask the secretary. Most times, the secretary is willing to reveal his or her information about finding Mr. Help's class.

Step C: Plan to Do the Exact Opposite from Your Past

- **When taking a math class, do the exact opposite from past negative experiences. Do the opposite of the past.**

In high school, I was *Quincy* the Quitter. I earned a "D" in my Calculus class, because I carried old patterns inside of me that contributed to my negative experiences. I only did homework at school, and I didn't study at home. I failed to do the nightly homework. I didn't go to the teacher for help. I thought that I would eventually understand the information on my own. That day never came!

I brought home my books from school, but I never opened them. Sure, I wanted to feel like a good student. But, I didn't have the follow-through to be a good student. Then, a breakthrough occurred. My grades began improving, when I started receiving tutoring from my teacher, Mr. Provencio.

During Winter Break, Mr. Provencio offered complementary math tutoring for our Calculus Advanced Placement Exams. He agreed to volunteer his time for anyone who wanted help. I signed up for the complementary tutoring. To my surprise, I was the only person who showed up.

While attending tutoring, I realized I needed to create good results in everything I set out to accomplish. I also learned that I needed to re-evaluate my past. Mr. Provencio encouraged me to do my homework every night. He showed me how to ask questions. Mr. Provencio taught me that it's okay to ask for help, because we're never able to understand everything ourselves 100 percent of the time.

It would be an amazing story if I said that I passed the Advanced Calculus class with "flying colors." But, that would be a lie. I failed the exam. However, during that Winter Break with Mr. Provencio, I learned that I needed to do the exact opposite of my past. This new *"Do the Opposite"* approach would guarantee me future success. Since I failed the Advanced Placement exam, I took another Calculus class at CSUN. My new *"Do the Opposite"* skills that I acquired allowed me to ace my first college math class. Plus, my *"Do the Opposite"* skills enabled me to help others as a math tutor.

As an exercise, make a list of all the actions you exhibited in your math classes, which allowed you to fail. Now, with that list, create a new list of all the opposite steps. Commit your new list to memory. What is your pattern? This new *"Do the Opposite"* approach will open your eyes to a new way of success.

Step D: Make Patient Mathematicians Your Friends

- **Study with patient people who are stronger in mathematics than you. This includes tutors and friends.**

Steps A-D all have a common theme: in each story, *Quincy* the Quitter found a person who was patient. If you want to be David the Determined, you must find a few people who can help you be successful. A lot of times, when we are exhibiting *Quincy* behaviors, we think that there is no one to show us how to do math. However, in every situation, there is always someone willing to help.

Think. There is always someone who is willing to help you. Despite the situation that you are experiencing, help is usually available. Math classes are no different. We can find people in math classes who are patient and more knowledgeable. If you find someone in a math class who has scored well, you can approach the person like this:

> Hi Sally, I can see that you seem to know the topic well. I know that you appear to understand the information. I do not understand certain math concepts now. But when I understand it, I do very well. Perhaps, you and I can study together for an hour next week. How does your schedule look?

Try this approach on a person in class. If your conversation with one person doesn't work, try another person. By three tries, you will find a person that's patient and willing to help you.

In summary, consider *Quincy's* versus David's Actions:

Quincy's Actions	*David's Actions*
• Doesn't do his homework	• Studies with a tutor.
• Assumes he will fail.	• Expects that he will succeed.
• Sits in the back of class.	• Sits with people in the front of class.
• Focuses on his failures.	• Focuses on his goals.
• Tries to study on his own.	• Finds people to study with.
• Keeps his book closed.	• Reads sections of the book and asks for verbal descriptions.
• Views math as intimidating.	• Takes his actions from other successful events and applies the same actions to math studying.
• Stops trying when his first attempt fails.	• Continues through his first success.

Okay, you have read the ways that David thinks. You have also read the way that David reprograms *Quincy's* brain. Lastly, you have discovered four TRIZ® action steps that help transform *Quincy's* brain neurons into David's determination.

Take these methods and try them for the next 30 days. Document your results. I am positive that if you stay on this plan for 30 days, you will be able to accomplish great feats. Now, let us see how we can convert *Donna* the Over-Doer into Sarah the Strategist.

TURN DONNA THE OVER-DOER INTO SARAH THE STRATEGIST

Chapter 7

Turn Donna the Over-Doer into Sarah the Strategist

Imagine a female leopard looking for food in an African jungle. When she locks eye on her next prey, she stops. She waits still and observes every step of the entity that she intends to conquer. She observes her meal moving left, right, then closer. The leopard observes her environment, and she takes advantage of the wind, trees, and shade. During this entire time, she is patient. She prepares her attack strategy. Then in one quick swoop, the leopard leaps and pounces with power. This jungle cat was successful, because she was patient, strategic and moved with certainty.

When we approach a math problem, patience, strategy and certainty are necessary. The solution process is identical to a jungle cat watching its prey. Every math problem is unique, and we have to watch closely to determine our attack strategy before any move is made. Without patience, strategy and certainty, we let our fears take over. And in turn, we become *Donna* the Over-Doer. In this chapter, we will help *Donna* divorce herself from compulsion through a three-step process; (1) *Donna* will discover the origin of her fear; (2) She will reprogram her dysfunctional thinking of needing to impress others; and (3) equip *Donna* with the ability to focus with strategy-based tools.

Step 1: Know *Donna's* Root Issue

Let us understand *Donna* better. We must know about the situations that motivate *Donna* to stress out. When Donna feels that she has to prove her knowledge, Donna's dysfunctional operations make her problems worse.

As described earlier, *Donna* is a person who does everything over the top. She meets in a study group, and memorizes all terms. *Donna* goes to the teacher for help. She makes flash cards. Yet, she is still confused with the various types of formulas and examples. Oftentimes, her actions lead to feelings of suffocation, and she panics. *Donna's* excessive habits are not offering her any solace. Consequently, all of her previous accomplishments are pushed to the back of her mind.

> *"When Donna feels that she has to prove her knowledge, Donna's dysfunctional operations make her problems worse."*

Donna believes her actions are "never good enough." In math, she feels like she can never "do things right." She is always wishing and hoping that her next moves will payoff. She is almost like an addicted gambler. The more she rolls the dice, the closer she thinks she is to winning the jackpot. In the process, *Donna* drains her emotional resources and self-confidence.

Secretly, *Donna* is seeking outside validation for her worth as a good student. It is easier for her to become distracted, rather than tackle all of her self-defeating actions. Behind closed doors, *Donna* has a fear that she will be rejected when she asks for support. With low self-esteem, *Donna* needs to feel like she has an outer support network. *Donna* believes the more she can show on homework/exams, and offer through class participation, she will retain and attract support. She believes her actions will prevent feelings of loneliness.

Donna sincerely fears being abandoned to sort through all of her accumulated thoughts about failing. She thinks that she can create "perfection," but she doesn't understand that it's impossible for anyone on earth to ever be perfect. If we do not find the root source of Donna's dissatisfaction, her excessive actions will lead to more compulsion. *Donna* needs help understanding why she suffocates herself with work and abandons her freedom.

Deep inside, *Donna* is a powerful person. However, *Donna* is accustomed to pleasing others. Thus, she refrains from releasing her true power in fear that she will scare off the people around her. In order to get to the root of Donna's issues, let's ask these basic questions. If you relate to Donna, then close your eyes (if possible), sit back, relax and voyage back in your memory:

At what age, did you first believe that your actions were not good enough? It can be within any event, math class, or unrelated scenario.

What happened when you tried to gain help from a person who you thought was a role model? What was the outcome?

How did you feel at that very moment? Explain those feelings on paper. Or, say it aloud.

Today, what part of that feeling reoccurs when you are sitting, and trying to learn in class?

When you thought about failing (or not understanding math), explain when you started to double-and triple-check your actions to avoid feeling anxious.

Now, describe a subsequent event that triggered your desire to work nonstop and fill up your free time.

Remember that event. Say aloud:

> "I cancel that moment in time. I forgive myself for choosing to accept fear in my past. I will no longer choose to be ruled by the fear that tries to creep into my mind. Fear, you are no longer welcomed into my life!"

Now, decide to change the outcome. Return to that moment in time. Pretend that you have nothing to fear. What would you say to that person? Say it.

Pretend that you chose to no longer act compulsive, and created free time for yourself. Describe how different the outcome would have been.

Open your eyes, and return to this moment. What comes to your mind?

There is good news: we can change *Donna's* self-sabotaging compulsive actions. But, in order to transform, *Donna* must ask herself a few motivational questions:

Questions Regarding *Donna's* Behavior:

In your youth, what activity did you genuinely enjoy doing?

Did you earn awards for the activities?

How did you set goals that led to your big awards?

How did you verify that you were working toward your awards?

How did your loved ones respond (or not) to your accomplishments?

Have you ever rewarded yourself for not being perfect?

Rewards can be additional personal time, a small present, clearing out your personal space, or a congratulatory letter to yourself. What opportunities exist within your future, where you can reward yourself?

Be Sarah the Strategist

In one sentence, Sarah uses strategy to transform her compulsion into freedom. Sarah exemplifies the type of person who has the ability to see the big picture, and knows that nothing in life is ever perfect. She is active, dynamic, courageous, and never afraid to take on a challenge. Sarah knows that calculated risks are part of her success. Sarah plans and figures out how to efficiently accomplish her goals. Sarah possesses confidence and sets realistic goals. Whereas, *Donna* claims to be capable of accomplishing a goal that is out of her reach! Before Sarah tackles intellectual assignments, she sorts through any chaotic emotions; then she grounds herself with planned actions.

> *"In one sentence, Sarah uses strategy to transform her compulsion into freedom."*

Sarah is not like Donna. Unlike Donna's need to please others, Sarah uses her resources to find the best possible solutions. She never obsessively ponders over things, nor re-thinks every possible response. Sarah independently handles the consequences of her actions with courage and without regret. She knows that she is not Wonder Woman™, and realizes that her methods are not perfect. However, she does the best that's humanly possible. Sarah understands that many things that matter are sometimes subtle and unseen. At times, Sarah becomes focused on these subtle cues by being alone, where she can be creative. Sarah combines flexibility with action, and without hesitation. Unlike Donna's perfectionist and no-rest attitude, Sarah accepts that countless things are out of her control. Therefore, Sarah accepts spontaneity and fully enjoys her experiences.

Step 2: Divorce Compulsion & Re-Marry Freedom

Remember, when *Donna* views math, she sees herself as "not good enough" to find the answer. As a result, *Donna* has mastered compulsive behavior. However, Sarah is a strategist, and knows how to plan to succeed. Thus, Sarah transforms Donna's thoughts, so *Donna* can become a strategist as well. With a mind reprogramming technique, Sarah blocks Donna's self-doubt, replacing each thought with confident, strategic, and creative confirmations that are recited aloud.

Sarah reprograms the way that *Donna* thinks:

- *Donna* thinks, "I can't memorize all the formulas. I have no idea what formula to use."

- Sarah Reprograms, "I will choose to know the key formulas. And, when they are commonly used. I will be able to solve problems with my accumulated knowledge."

- *Donna* thinks, "I must impress the teacher, and get an "A.""

- Sarah Reprograms, "I will answer the questions asked of me to the best of my ability."

- *Donna* thinks, "I don't know what I am supposed to do."

- Sarah Reprograms, "I will watch others to find out what I am not supposed to do."

- *Donna* thinks, "I am smarter in English, writing and other creative subjects."

- Sarah Reprograms, "With a strategic effort, I can be just as brilliant in math, as I am in English and other subjects."

- *Donna* thinks, "I have to do well in my life in order to be accepted."

- Sarah Reprograms, "I can accept myself, and I was perfectly created."

- *Donna* thinks, "I only have to take the math requirement, and then I can stop."

- Sarah Reprograms, "I will integrate my math knowledge to solve real-life problems, so I can have an edge over any future competition."

Sarah's Daily Affirmations:

I am enough.

Imperfection is the only way to become stronger.

Now, I will embrace my past failures, and be proud that I can re-learn.

My plans do not consume my time, because I am flexible.

I will accept advice from math experts. All others' advice will be secondary.

I will not follow what others are doing in math. I will create my own success plan.

I will create a strategy that includes rest.

Step 3: Strategy = Success

A strategy does not fall out of the sky. It is a plan that helps us obtain our goals. Strategies are formed by using the resources around us. Such tactics can be creative, and are interchangeable at any time. Your strategy is a tool to help achieve your goals efficiently. Specifically in math, a strategy is an action plan that allows you to attack, and completely solve math problems.

The route to creating a strategy is simple – ask the right questions. What is your goal? What direction can you take *now* to move toward your goal? What meaningful difference will you provide once your math goal is accomplished? The process to creating a strategy is first envisioning your end goal. Next, it consists of using the resources that you have available. And, it is applying your plan, and watching if it works. If it doesn't, you can adjust your plan to bypass any obstacle.

There is no such thing as a perfect strategy. However, we can use our creativity to manifest our goal. Then, we can use our resources to transform our vision into reality. Along the way, we can make changes. Eventually, we will accomplish the desired goal. However, we must be patient within this process.

Use the 5 S.O.S. = 5 Strategies of Success

Strategy 1: Declare War On Distraction

Have you ever started to study, then took an incoming phone call? Or, have you sat down to study, then, thought about cleaning your laundry? Ever wanted to study, but received a distracting text message? You cannot accomplish your goals if you are distracted. We cannot create an effective strategy for our success, unless we identify all of our distractions. It's also vital to identify each distraction and inwardly declare war.

Thus, when your time runs out, you can always look back to gauge how your time was wasted. When we declare war on distraction, we are foreseeing potential threats, and eliminating them, before they foil our plans. In the cases above, we can remove distractions by turning off our phones and emails. Plus, we can move away from piles of dirty laundry.

Amidst the turmoil of math confusion, keep your presence of mind: despite the bewildering circumstances. When I was in graduate school, the subjects were rather difficult to understand. As a result, I found myself trying to study at home. I was frustrated. Consequently, I found anything and everything to distract me. I baked cookies, and I washed laundry when I needed to do so. I eventually talked on the phone. In each case, I found myself returning back to my math problems. I was trying to memorize terms and concepts without effectively studying. People with a "*Donna* personality" will get distracted during the act of studying, without understanding that effective studying requires timed breaks.

When we learn to detach ourselves from distraction, we also must strategize how to operate with focus. Select a set time, daily to study math. Is it from 3-5 p.m.? Or, is it 6-8 p.m.? Whatever time that you choose, you can find a way to exit the distraction by remembering your #1 priority: to study math, and then to find time for your sanity.

Strategy 2: Set Realistic, Definable Math Goals

Many times, we can be in a math class and think that we must have 100 percent correct answers. This is the cruelest expectation that you can place on yourself! Part of successful, strategy-building includes incremental goals. For example, if you have to study eight sections of math, prioritize the order of sections that you must understand first. Give yourself a set time to cover that section! Then, move on to the next. If you do not finish it, go back to it later. The concept is not going anywhere.

Another way to define clear goals is to trade space for time. By this, I mean that the *Donna* personalities typically fill up their physical space with items they think will prepare them for success. Even though this sounds counterproductive, we must give away or throw out things that we do not need. Get rid of items such as old emails, clothes, and filed papers – then you can free your mind to focus. By spending 10 minutes a day throwing out things, we gain an hour of mental clarity. Start with a set time dedicated to making space for positive change, and new thoughts.

Next, before a class starts, plan to score lower than 100 percent on your first test. Yes, you read correctly. For example, if there is an opportunity to take five tests during a semester, you can make Test 1's goal to be a score of 90 percent. A second test can gain 92 percent. And so on. Allow yourself to make mistakes, but also permit yourself to know that you can acquire a high score at the end of the class. As a college educator for more than a decade, I will share with you a little secret that teachers often fail to share with their students. If a teacher notices significant improvement by the end of the class, the teacher is more apt to give a higher grade. Most teachers look for improvement over the semester or quarter, rather than perfect scores.

> *"If a teacher notices significant improvement by the end of the class, the teacher is more apt to give a higher grade."*

Strategy 3: Realize "A's" Does Not Equal Smart

I have been teaching for more than 10 years. And, to my surprise, "'Weaker students' or people suffering from math innumeracy are capable of earning straight 'A's." You may wonder, "How on earth is this possible?" Students that get straight "A's" have mastered how to express math on tests. Though, they may not be able to explain or use the theory behind many calculations. These students are able to regurgitate the math that teachers want on tests. Conversely, some of the smartest students in the class may earn "D's," simply because they lost their focus.

> *"Your grade has nothing to do with your intelligence. "*

The point: just because you did not get an "A" before, doesn't mean that you cannot get an "A" now. Stop associating educational success with your intelligence level. Your grade has nothing to do with your intellect.

These two items are unrelated. The great statistician and philosopher W. Edwards Deming stated, "Tests are not a measure of intelligence, but simply, a measure of how effective a teaching method is received," (see Deming's book, The New Economics).

The older we are, the easier our brain can master harder topics. Our frontal brain lobes mature fully by the age of 25. Therefore, the brain is more capable of logical decision-making. Car insurance companies know this fact as well. At the age of 25, most drivers' insurance rates decrease. Why? You may ask. The answer: *people are able to use their brain more logically to avoid accidents.* Numerous times, I have witnessed students falsely believe that they will fail math, because they failed to previously grasp a math topic. This is the worst assumption that anyone can make.

RYANE

One day, I tried to tutor Ryane, a student in a Pre-Algebra class. She was 12-years-old, and struggling with basic Pre-Algebra concepts. As I taught the concepts, I saw her hesitate and her body pulling back. Then, she grew quiet. I asked what was wrong. At that moment, Ryane burst into tears. Sobbing, Ryane said she felt ashamed because at 12-years-old, she still didn't understand multiplication. She never learned it in third grade, and this was devastating for her! I hugged Ryane, reassuring her that she was fine. I explained that now she was old enough to relearn the concept quicker than if she were in third grade. Then, I proceeded to show her how to multiply. As Ryane watched the concepts unfold in front of her eyes, you could see her spirit come alive. Ryane's tears stopped, and a smile came upon her face. She became amazed and yelled out, "That's it! That's all that I have to do? I thought it was much harder than that!" Her epiphany came! Her fear was gone, because she realized a few basic principles: It's perfectly okay to relearn concepts despite our past. We are never too old to learn. Our past may not be perfect, but our future can be.

Strategy 4: Demand to See "The Big Picture" in Math

Have you ever tried to understand why you used a certain formula in mathematics? If you have, then perhaps you were missing the big picture in mathematics. The "Big Picture" answers *"Why?"* instead of *"How?"*

We need to ask, *"Why"* in mathematics too. We need to question why we would want to use a certain approach versus another. We want to understand why the concept and procedure exists. When we gain answers to the "Why" questions, we can look at math problems with a big picture view.

Example:

We want to find the distance in two cases:

1)   Find the distance between two points (2,0) and (8,0).

2)   Find the distance a bike will travel at 2 miles per hour for 8 hours.

These problems are similar. We must find the distance. In the first case, we must find the distance between two points on the x-axis. In the second case, we must find the distance between where the bike starts, versus where it stops. Now, *Donna* would panic. She would think that she would have to memorize the two formulas in each case.

Logically, we would like to know why these two distance problems are different. In the first case, we are dealing with points that are stationary. While in the second case, we are dealing with a moving point. We can figure out the distance using logic and without formulas. In Case 1, since the points are not moving, we use the distance formula, $d = \sqrt{(x_2 - x_1)^2 + (y_2 - y_1)^2}$. In Case 2, since the points are moving, we use the distance formula, $d=rt$ to find how far the bike travels.

We use this formula when the points are not moving. This is the distance formula:

$$d = \sqrt{(x_2 - x_1)^2 + (y_2 - y_1)^2}$$

Where (x_2, y_2) and (x_1, y_1) are two points in the Cartesian Coordinate System.

In our example:

$$x_2 = 8,\ x_1 = 2,\ y_2 = 0,\ y_1 = 0$$

$$d = \sqrt{(8-2)^2 + (0-0)^2}$$

$$d = \sqrt{(6)^2}$$

$$d = \sqrt{36}$$

$$d = 6$$

So the answer is 6 units.

In Case 2, this is the distance formula, since the point is moving:

$$d = rt$$

Where r = speed of the object, and t = time the object travels.

In our case, r =2 mph and t = 8 hours

$$d = rt$$

$$d = (2mph)(8hrs)$$

$$d = 16 \text{ miles}$$

These examples are shown, because we would have to know why a formula is chosen, rather than memorizing it. When we ask *"Why"* instead of *"How"*, we then can think about the logic behind a problem. When the logic is discovered, the formula becomes clear.

Strategy 5: Demand Positive Feedback from Teachers

Teachers are notorious for telling you where you failed. But how many times have teachers explained to you the areas where you succeeded? Well, it's time to receive that information. To become a Sarah personality, we must ask for the steps that will help us succeed. Instead of penalizing ourselves whenever we need help, we must do the opposite.

Now, you may wonder how you can convince your teacher to offer you good information. One way is to penetrate their mind with communication. Lure your teacher into communication by allowing the teacher to delve inside your mind. Ask the instructors to indicate if they can understand your logic. Request the teacher to identify when you created a step correctly, and not incorrectly. You must know the ways to create success, and not focus on your math failure. Visit during the teacher's office hours.

Ask, "Can you point out, in detail where I applied concepts correctly?"

This question does three things. First, the question shows you the areas that you understand well. In turn, you can build your confidence. Secondly, the question requires that the teacher explain the process between problem types more thoroughly. It opens a two-way communication, to compare and contrast problems with the same concept. Thirdly, the question sets the precedence for continuous positive, constructive feedback throughout the course. You are building a personal rapport with your instructor, as well as holding the teacher accountable for teaching.

Donna's Actions	Sarah's Actions
• Obsessively ponders over "What if scenarios."	• Naturally find the best possible response.
• Thinks that her perfectionist (and no-rest Wonder Woman™ attitude) can help her ace math.	• Accepts that many things are out of her control, but she will do her best.
• Focuses on memorizing formulas.	• Sees the "big picture," and understands why formulas are used.
• Afraid that she will do something wrong and fail.	• Takes calculated risks after she is fully prepared.
• Seeks outside validation from others regarding her knowledge.	• Picks 1 or 2 mentors and only asks these individuals for insight. No others' opinions count.
• Fills up her physical space with items that she thinks she will need.	• Regularly dumps out things she doesn't need.
• Thinks that an "A" = Smart.	• Knows that unskillful people can get "A's" too.
• Stays quiet and doesn't ask the teacher for help.	• Demands that the teacher show what she did well, and requests specific tips on problems.
• Studies day and night without rest.	• Schedules one day of 100 percent "Me" time.

TURN SAMUEL THE STRUGGLER INTO IVAN THE INNOVATOR

CHAPTER 8

TURN SAMUEL THE STRUGGLER INTO IVAN THE INNOVATOR

He did not speak until age 3. In his youth, he could not express himself in a written language, and found math difficult to explain. Many of his teachers thought that he was simple-minded. One day, his teachers noticed he could achieve great things simply by visualizing rather than speaking. Through his advanced visualization, he eventually revolutionized modern physics and discovered the theory of relativity – all of which was created in his spare time. Even as an adult, he found it laborious to explain himself. As a *Samuel personality,* this struggling student transformed himself into an innovator. This amazing man was Albert Einstein.

Albert Einstein was one of the first recorded cases of who I call *Samuel* the Struggler, but he eventually capitalized on his own thinking power to become Ivan the Innovator. Frequently, individuals who struggle to learn, both male and female, are known as "*Samuels.*" This student is commonly perceived to be a "slow learner." Many times, these "*Samuel* personalities" don't receive the same respect as other students. In traditional learning environments, Samuel personalities are sometimes labeled as having a learning disability.

However, no one has taught these students effective ways to use their brain in order to learn quickly. As a result, *Samuel* students are at a disadvantage, because they do not know how to capitalize on their unique thinking power.

> *"In traditional learning environments, Samuel personalities are sometimes labeled as having a learning disability. "*

In Chapter Four, we read that each person's brain is different: in turn, our brains transmit information distinctive ways. With unique DNA, which forms neuron pathways, we each have a unique brain operation with unique thoughts. Einstein's formula, $E = mc^2$, also represents the billions of potential neuron energy pathways that can be used to transform thought into action. Our thoughts have energy and power. However, if we do not capitalize on our brain's unique, powerful operation, we may do more harm instead of good. In this chapter, we will concentrate on three aspects that will change *Samuel* into Ivan. First of all, *Samuel* must understand that his brain works differently than others. Secondly, he needs to fuel his brain with the right foods. Third, *Samuel* must give himself enough time to solve problems, then he can transform into the great Ivan-the-Innovator!

Step 1: Know *Samuel's* Root Issue

As we saw before, *Samuel* is a genius in disguise. With his unique brain, *Samuel* doesn't know how to show his own intelligence to a world, which uses a different logic. *Samuel* is a person with *Special Needs,* who is neither weird nor odd. *Samuel* wants what everyone wants: to be accepted. But first, he must learn to accept himself. His main challenge is to be understood by others. *Samuel* ought to learn how his brain operates, so he can accept his brain. Next, *Samuel* should discover the way that others communicate. By taking the steps in this chapter, he will be an effective problem-solver known as "Ivan the Innovator."

For *Samuel* the Struggler:

Circle the "weaknesses" that has affected you the most:

Listening Comprehension

Learning Oral Language

Reading Skills

Written Skills

Social Relationships

Poor Eyesight

Scrambled Note Taking

Inverted Numbers or Symbols

General Health Issues

Blood Sugar Levels

Poor Hearing

Short Attention Span

Hormone Problems

How has one or several of these weaknesses hindered you during your learning process?

When did you first discover this difficulty? Now, what did you do to minimize the fact that you were different?

At what age did you experience your first major discrimination because you were different? It can be any event, math or unrelated to math.

What happened when you tried to gain help in that situation? What was the outcome?

Explain three, self-defeating thoughts you created at that moment that impacted the way you thought about your own capabilities.

How did you feel at that very moment? Explain those feelings on paper or say it aloud.

When did that feeling reoccur? Was it in a math class, during math lessons or within your math homework?

When you thought about being different in learning math, how did you cope with those feelings?

Pinpoint a distinct, subsequent incident, when you decided to ignore the teacher or become angry toward the teacher for not understanding you.

Remember that moment in time. Now say aloud:

> "I cancel that moment in time. I forgive myself for choosing to accept fear in my past. I will no longer choose to be ruled by the fear that tries to creep into my mind. Fear, you are no longer welcomed into my life."

Now, decide to change the outcome. Return to that moment in time. Pretend that you have nothing to fear. What would you say to that person? Say it.

Pretend that you refuse to get angry or discouraged for not being understood. Describe how the outcome would have changed.

Next, look at the situation with new eyes. Pretend that you chose to celebrate how your brain works. What would you say to your teacher or tutors so they could understand how you think?

Open your eyes, and return to this moment. What comes to your mind?

Become Ivan the Innovator

Ivan has the ability to think and create new inventions and services that people need. As business people say, "Ivan thinks, 'out of the box.'" Ivan thinks beyond himself, and uses life's physical restrictions as a compass to find the right path toward overall success. For example, Ivan may find a way to use cell phone technology to communicate with ATMs before that technology is readily available. Or, an Ivan personality may find a way to use the Internet to call someone before that technology is used by the average public. Ivan also responds to new opportunities before others can realize the need.

"He creates new ways when no one else's brain can visualize it."

For instance, Ivan may see that people desire to use facebook© safely while driving. Thus, he creates a facebook©, hands-free status application system to use in cars. Ivan improves old techniques that do not work, while inventing new solutions that everyone can use. Ivan's gift is always to improve the method by which things work. Ivan is dynamic, and he acts as a leader, and facilitator for change. He creates new ways when no one else's brain can visualize it.

In math, Ivan understands how he learns, and applies his knowledge to his math education. Ivan sees that he has all the resources he needs to make positive changes in his math comprehension. He enjoys using his unique abilities to understand math. Then, Ivan improves the math by finding new approaches to solve problems. More importantly, Ivan finds ways to re-explain math to others in their language, so they can understand his logic!

A True-Life *Samuel* Named Angelica

Angelica was a student in my Beginning Algebra class. She was a quiet person, and rather shy until she sat with her friends. One day, a particular group of students became unusually loud during a class assignment. Without pointing out anyone, I reminded the class to focus on the math. However, Angelica and two other students stood out to me. I knew I had to address their excessive talking after class. I asked Angelica and her buddies to remain after the class ended.

I said:

> I am pulling you to the side to speak directly to you three. In class, I am committed to helping each of you understand this work. I will answer any question that you may have, and I will also stay after class to give you coaching. In return, I request that you stay focused, and not let fear distract you. I have high expectations, and I am here to help you do ex-

tremely well in this class…that is…if you work with me. And, don't talk!

As I watched the fear transform to respect in their return gaze, I witnessed each of them understand the class expectations. Then, a few weeks later, an epiphany hit me like a ton on bricks! Angelica was a *Samuel*! Angelica *tried to hide her struggle by talking too much*. Through an assignment, it was confirmed that Angelica was a *Samuel* the Struggler. I asked about when she first discovered that she needed math help and found math difficulty. Angelica responded with her story:

> When I was in second grade, we were told to work on some math problems, (like dividing) with a group of teacher's assistants. In my group, there were about two other people besides me. The teacher's assistant did not help each of us as a group. But, she helped us individually. I was the last person she came to instruct in my group. I could not understand how to divide. A simple problem like "$8 \div 2$" was given as our assignment. I could not figure out what this meant. When the teacher's assistant was trying to explain, she just told me to simplify by a using a number that can go into both of these numbers. I had no idea what she was talking about. So I freaked out, and I guessed the number "4!" She looked at me and said, "No." I still did not understand her. She did not tell me how to divide. So, for the longest time, I had trouble with dividing. I had trouble with math in general because I was afraid to ask questions.

After reading Angelica's experience, I realized that Angelica was a perfect *Samuel* the Struggler. *Samuel* the Strugglers usually had teachers who gave up on them. As a result, Angelica lived in fear that her teacher will not understand her work. Or, he would not explain concepts to her because she was assumed to be "slow." Consequently, I began to work with Angelica so she could divorce the stigma that she unknowingly accepted.

Divorce Stigma & Re-Marry Respect

Motivational speakers do it. Inventors do it. World leaders do it. Athletes do it. CEOs do it. All triumphant people do it! They move past fear, and envision their success before it happens! They program their minds to achieve greatness, instead of falling short due to stigma. In order to change the way that the world views us, we must change the way that we picture ourselves. There is a famous saying: *"You must treat yourself the way that you want others to treat you."* This statement is true in all of our lives, especially *Samuel's*.

If *Samuel* continues to rely on others to respect his thoughts, he will forever fall short of reaching his full potential. However, Ivan is the polar opposite of *Samuel*: Ivan knows that he must respect how his own brain works first, before he expects others to respect him. Thus, others will learn to respect Ivan's technique for communicating and receiving information. To compare, we will investigate Ivan's method that changes *Samuel's* perception.

In this case, *Samuel* can reprogram his mind to think different thoughts: Ivan corrects *Samuel's* negative thinking by capitalizing on his special uniqueness.

- *Samuel* thinks, "Wait! This math is going too fast."

- Ivan Reprograms, "I will ask the teacher to explain the math at a slower pace."

- *Samuel* thinks, "I feel so scatter-brained … I can't absorb what the teacher says."

- Ivan Reprograms, "I will focus my brain to see the main purpose of this calculation."

- *Samuel* thinks, "I have trouble focusing in math. I can't help but think of other things."

- Ivan Reprograms, "It is OK to daydream at specific times. This is how I gain my brilliance. I will attempt to focus for 10 minutes at-a-time. Then, I will take a 1-minute break. I will keep doing this, until I can finish my work."

- *Samuel* thinks, "I feel inferior to other people, because I am different."

- Ivan Reprograms, "I feel proud because I am made exactly the way that I am supposed to be. I can learn anything when I am focused."

- *Samuel* thinks, "My math teacher thinks that I am stupid, because she can't read my handwriting."

- Ivan Reprograms, "If I have to draw pictures and arrows so the teacher can understand me, I will do it. I will guarantee that my logic is always understood."

- *Samuel* thinks, "I sometimes transpose numbers, and I write numbers in the wrong order."

- Ivan Reprograms, "I will have someone to double-check my work, before I submit it. I will also gain help from the learning center on my campus. They can help me find the tools to communicate with others clearly."

Ivan's Daily Affirmations:

My brain is special, and I am smart.

Many people must learn the way I understand things.

I will learn the way that others communicate, so they can appreciate my thoughts.

I will seek all the help I need.

There is always someone willing to help me.

I am worth the time and explanation.

Transforming Angelica into Ivan

I worked with Angelica to accomplish three basic steps. I wanted her to help herself do well in mathematics. First, I helped her to recognize that "placement is the key." Let me better explain. After I pulled Angelica aside, I indicated that she needed to sit near the front of the class. If she sat in the front, she is less apt to distract herself from the teacher, who is directly in front of her. Next, I told Angelica that it was OK for her to ask me questions. I told her that I expected her to not to understand my approach at first. But, also so that she would eventually understand. I also told Angelica that she must ask teachers questions outside of the lecture. I explained that I expected her to eventually execute the concept by the end of the class. I indicated that I would answer her questions, both large and small, provided that she committed herself to gaining help. Lastly, I emphasized to Angelica that she needed to focus on her mission. I assigned her to class groups, where people would work at a pace that would allow her to digest the information thoroughly, and within a set time. Angelica also brought her papers to me so she could write the math in a way that anyone could understand. She successfully completed the course with a "B" grade.

Are You Killing Your Brain Power?

When I was working as a rocket scientist, I ate four donuts during a briefing one morning. Then, we went into the Launch Control Room, where I quickly gulped down two cups of coffee. This insane sugar rush helped me for a total of four hours. Then, I started spiraling downward, becoming unbelievably tired for the next few hours. I struggled to keep my eyes wide open and alert. However, this was a HUGE problem, because I was supposed to be alert in order to prevent launch pad explosions! Gradually, my speech became somewhat slower. Appallingly, it took me a few minutes, instead of a few seconds, to solve problems. Little did I know I was essentially choking off the oxygen supply to my brain!

Sometimes, you unknowingly hinder yourself from learning, because you choke your brain of oxygen by placing harmful substances in your system. Harmful substances are more than just the typical drugs and alcohol. Substances include sugar, processed foods and soda. Our intake can either starve our brain of oxygen or add oxygen to it. The trick is adding as much oxygen as possible!

After my sugar-caffeine-high-experience, I suffered severe, allergic reactions for the next few days. I know that two donuts and coffee is nothing for the average person to eat. But, when you are a *Samuel*-type personality, you can be overly-sensitive to ordinary things that others can handle. My sleep pattern was greatly disturbed. I would fall asleep early, and then wake up in the middle of the night. For the next few days, I experienced a "fog-like" mist over my thoughts. Following these abnormal events, I learned my brain was sensitive to sugar and caffeine. I also discovered I was allergic to gluten and yeast found in donuts.

Processed foods make our body work harder to digest food. More importantly, preservatives in these foods act as poison to the body. Therefore, the liver tries to isolate preservatives by creating mucus in the body. Simultaneously, the liver sends messages to the hypothalamus in the brain to adjust hormone levels to avoid complete digestion. The brain is tied up. In turn, your brain slows in its operation in learning.

While also observing the pattern of my intake, I noticed my brain operated faster when I oxygenated my system with vegetables and whole foods. I noticed that when I polluted my system with sugar, milk, red meat and stress, my brain appears to take longer to function. During such circumstances, it feels as if I have a learning disability. It is through such experiences that I realized we have power over our own learning capacity.

My goal is to share the **Top Brain Killers** we must remove from our lives in order for our brains to function faster. I will offer replacements to help maximize our brain's natural capabilities. The following list is based on a simple idea. Some variables remove oxygen, thus fog our brain. While other things, add oxygen to our body, which, in turn, feeds our brain. When our brains are fed, we can be innovators, capitalizing on our own thinking power. Avoid these pitfalls, if you can:

Top Brain Killers:

Stop Hitting the S.A.C. – Soda, Alcohol & Coffee

Caffeine in coffee and sugar in soda greatly impairs the message transfer process of neuron-to-neuron within the brain. Therefore, information gets trapped in one area of the brain, preventing quick movement to the next. Soda, coffee and alcohol are processed by the kidneys in the body. The kidneys can become strained from these substances, causing the liver and pancreas to alter the hormone secretion in your body to compensate for the high levels of sugar and caffeine. The brain readjusts its hormone secretion: and the pituitary gland is affected. Less neuron-to-neuron information transfer occurs in the brain. Essentially, your brain short-circuits.

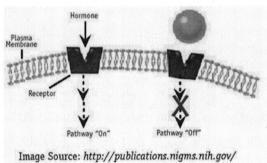

Image Source: *http://publications.nigms.nih.gov/ chemhealth/images/ch4_party.gif*

Alcohol, similar to tobacco use, also creates free radicals that damage the message transfer process. Abuse of alcohol not only impairs the message transfer process in the same manner as smoking, but also causes severely-low oxygen levels in the bloodstream. This leads to the destruction of brain cells.

Part with Processed Foods

I remember being 19, stressed out and studying for a test one night. I was starving, and ate a 16-oz container of ready-made potato salad! I experienced a horrific migraine for three days, and couldn't study anything. Guess why?

Think about a juicy fast food burger, a delicious frozen pizza, and tasty bag of chips or mouth-watering cookies. What do they have in common? The answer is MSG or Monosodium Glutamate. MSG is a food preservative, and a salt of the amino acid "aka" glutamate. It is a preservative reportedly known to cause adverse effects such as headaches, sweating, and nausea, numbness in the face, burning sensation, chest pain and weakness. Unfortunately, a lot of those fast food burgers, pizza, and chips are typically not made with real food. Instead, they're loaded with preservatives.

Check your labels, preservatives can be found in nearly everything! You'll find preservatives in canned soups to flavored potato chips; to quick-cooking noodles and canned gravy; to frozen prepared meals; to salad dressings, especially the "healthy, low fat" ones. Items that don't contain MSG often include something called Hydrolyzed Vegetable Protein, which is just another name for MSG.

There are more preservatives than MSG. The additive *butylated hydroxyanisole* preserves food, allowing it to stay on the shelves longer. While this preservation process may have been a good idea for pilgrims on the *Mayflower*, today preservatives in processed foods are responsible for a slew of problems such as Type II diabetes in youth, thyroid balance issues and excess weight. Chemical preservatives such as benzoic acid, sulfur dioxide, and sodium nitrite have long been used as an additive in foods. Sulfur dioxide has been known to irritate the bronchial tubes in our lungs and reduce oxygen to the brain.

Now, we do not have to quit "cold turkey" on processed food. However, we can start to limit our intake of such foods. As for me, I could never eat that type of potato salad again, because the *MSG* preservative caused my migraine. My body, as well as those considered having *Samuel*-type personalities, tend to be extremely intolerant of food preservatives!

Split from Stress and the People Who Carry It!

Have you ever been taught by a really bad teacher who wanted to fail you? Or, have you worked for a supervisor who wanted to fire you? Ever experienced a bad romantic break up? Each situation has the same common theme: STRESS! It causes our bodies to pump out increased levels of the hormone cortisol. Excessive cortisol levels can lead to the destruction of neurons in the hippocampus (the learning and memory center of the brain). Stress hinders the frontal brain lobes from working. Consequently, stress activates the same reptilian portion of the brain as mathaphobia®. Therefore, information gets trapped in one area, and it doesn't move to the next area of the brain. The following chart helps to summarize the replacements that we need.

Turn *Samuel* into Ivan Using Actions Steps:

Remove	Replace
• Soda/Coffee/Alcohol/Tobacco Caffeine in coffee and sugar in soda impair the message transfer process (neuron-to-neuron) within the brain. Tobacco use generates free radicals that impair the message transfer process. Abuse of alcohol not only impairs the message transfer process in the same manner as smoking, but also causes the destruction of brain cells.	**• Spring and Alkaline Water.** The ph-level in spring and alkaline-based water allows the body to better metabolize oxygen into the bloodstream. The electrolytes found in some sports drinks and waters allow the body to metabolize oxygen as well. Whole food vegetable juices also add chlorophyll to the body, which increases brain function. Green teas without caffeine are also good for the brain. Gum helps people who have a hand-to-mouth complex when under stress. Though gum would eventually have to be removed in time as well.

• **Remove Stress**	• **Meditation and Prayer**
Stress causes your body to pump out increased levels of the hormone cortisol. Excessive cortisol levels can lead to the destruction of neurons in the hippocampus (the learning and memory center of the brain). Stress hinders the frontal brain lobes from working. As a result, stress activates the same reptilian portion of the brain as mathaphobia®.	Three ways appear to stop the cortisol secretion: exercise, laughter and prayer. Exercise begins to re-establish the hormone regulation in the body, which stops cortisol secretion. In a study done with laughter, patients who laughed daily healed their ailments faster than a group without exposure to laughter. And, studies show that prayer and meditation are effective in reducing cortisol secretion.
• **Sedentary Lifestyle**	• **Active Lifestyle**
Sitting too long in one place does not allow blood to flow to our brains efficiently. If you let a car sit in a garage, you will have to start the car to get the fluids moving again. The car fails to operate at its optimum level until it is heated. Your body is the same. When you move your body, blood flows to your brain.	Studies of senior citizens who walk regularly showed significant improvement in memory skills compared to sedentary elderly people. Walking also improved their learning ability, concentration, and abstract reasoning. Stroke risk was cut by 57 percent in people who walked as little as 20 minutes per day. Plus, Vitamin D from the sun can increase brain function. Thus, walking in the sun can make you smarter.
• **Processed Foods and Additives**	• **Foods for your Blood Types**
Processed foods make your body work harder to digest food. And, the preservatives in these foods serve as poison to the body. In turn, your brain slows in its operation in learning. This list includes artificial colors, aspartame, BHA (butylated hydroxyanisole), MSG (monosodium glutamate), OLESTRA (Olean), sodium nitrates, sterol esters, sucralose, and sulfites. Using certain preservatives has been known to cause mild allergic reaction in sensitive people, including *Samuel*-the-Struggler students. Some animal studies suggest they create a small risk of cancer or tumors; and possible link with hyperactivity and learning disabilities.	Good diet and exercise habits reduce the likelihood you'll develop high blood pressure, diabetes and heart disease. Healthy eating according to your blood type can increase proper digestion, which leads to better oxygen flow to your brain.

1. Special Services = Special Results: Go to the Special Learning Testing Center on campus to know, in detail, how your brain works.

Sometimes, students will see numbers that are inverted. Or, students may find it difficult to write words on the paper, while listening to a lecture. In some cases, students are unable to see the board due to poor eyesight. At times, students face difficulties retaining information due to memory problems. Despite any learning challenges, there is always hope and assistance available.

> *"Despite any learning challenges, there is always hope and assistance available."*

From the 1950's through the 1990's, there was a stigma attached to individuals who had "special needs." Such students were sent through a program with "special services." Oftentimes, parents avoided having their kids "tracked" through these centers because school records would indicate the students had a "learning disability." As a result, parents would deny that their children needed special services. Sadly, thousands of children missed out on receiving one-on-one assistance for college-bound success! It is my belief that our national dropout rate is primarily composed of people who never received special education services! Thankfully, that time is far behind us. Special Services is a standard resource program provided by most colleges and universities for anyone who needs help.

Special Services is the program on campus that provides resources to students with learning disabilities. Each semester, thousands of students with challenges receive a wide range of support services not provided by other departments on campus. This "one-stop" approach creates a more user-friendly atmosphere for each student to capitalize on their natural brain capabilities. Most special services provide the following list of resources:

o Counseling

o Priority Registration

o English/Math Placement Tests

o Learning Disability Verification Testing

o Interpreting

o Note taking

o Adapted Computers

o Special Orientation

o Testing Accommodations

o Special Classes

o Mobility Assistance

o Wheelchair Loan

o On-Campus Transportation

o Brailler & Braille Printer

o Print Enlarger

o On/Off Campus Liaison

o Tutoring

Oftentimes, students receive Myers-Briggs® personality assessments and brain processing tests. Coaching is also available for students who need help. As a result, students discover they have help, and are not alone in their journey toward success.

2. Stop Limiting Your Time: Find ways to take tests and do assignments with ample time, so your mind continues to function.

I had a student in one of my Beginning Algebra classes who understood the lecture, but had other challenges. She could finish each math problem correctly, and answer most problems in class. However, when she sat down to take a test, she would freeze up. This poor student was experiencing dreadful test anxiety. Each time, she sat down to take an exam, her brain would go blank. Of course, she was frustrated over her performance.

Unbeknownst to her, anxiety was triggered by the test's "limited time." For many, restraints on tests end up actually being unfair, and often reduce intellectual capability. Unfortunately, timed tests aren't going away any time soon. We must learn to play the educational "game" to our advantage. Thus, to play the game, we must request more time.

Did you know that a commonly requested accommodation on College Board tests is extended time? When requesting it, schools are asked to indicate the specific subject area(s) in which extended time is needed (reading, written expression, mathematical calculations and speaking), as well as the amount of time the student needs. Students who request more than 100 percent extended time must provide documentation of their disabilities, and their need for accommodations for the College Board's review. SAT tests and other standardized tests accept this form of time extension. In order to quality for time extensions, most qualifying individuals visit their Special Services center on campus to gain the appropriate paperwork. (See STEP 1 above).

3. Follow the book, Eat Right 4 Your Blood Type by Peter J. D'Adamo. Brain function correlates to our food intake.

According to naturopath Peter J. D'Adamo, N.D., in his book Eat Right 4 Your Type, people who eat according to their blood type maximize the amount of oxygen traveling to their brain. Four basic blood types exist: O, A, B, and AB. His research into anthropology, medical history, and genetics led him to conclude that, "Blood types are responsible for how we digest or fail to digest our food." It is suggested that indigested food serves as toxins in our bloodstream and reduces our brain's optimum performance. In the book, food nutrition guidelines are given to help people eat according to their blood type. In eating per these guidelines, we have a better chance for thinking quickly and learning at a faster rate. The following blood types are given:

Type O - People with Type O blood fare best on intense physical exercise and animal proteins, and less well on dairy products and grains, says Dr. D'Adamo. The leading reason for weight gain among Type O's is the gluten found in wheat products and, to a lesser extent, lentils, corn, kidney beans, and cabbage, Dr. D'Adamo explains. In his book, the ideal exercises for Type O's include aerobics, martial arts, contact sports, and running (page 93).

Type A - Those with blood Type A, however, are more naturally suited to a vegetarian diet and foods that are fresh, pure, and organic. As Type A's are predisposed to heart disease, cancer, and diabetes, "I can't emphasize how critical this dietary adjustment can be to the sensitive immune system of Type A," suggests Dr. D'Adamo in Eat Right 4 Your Blood Type. He says Type A's prefer calming, centering exercise, such as yoga and Tai Chi (page 139).

Type B - Type B's have a strong immune system and a tolerant digestive system and tend to resist many of the severe chronic degenerative illnesses, or at least survive them better than the other blood types. Type B's do best with moderate physical exercise requiring mental balance, such as hiking, cycling, tennis, and swimming (page 181).

Type AB - Blood Type AB, the most recent in terms of evolution of the four groups, and an amalgam of Types A and B, is the most biologically complex. For this group, a combination of the exercises for Types A and B works best, says Dr. D'Adamo in his book on page 222.

While we cannot change our blood type, we can use knowledge about its nature to implement a dietary plan biologically-suited to our makeup. "Most of my patients experience some results within two weeks of starting the diet plan: increased energy, weight loss, a lessening of digestive complaints, and improvement of chronic conditions such as asthma, headaches, and heartburn," explains Dr. D'Adamo on his website *www.damo.com/program.htm.*

4. Write math in sequential steps, and not all over the page.

In the first year that I taught at the college level, I met David. He was a *Samuel*, but did not understand this aspect about his mind. On the first test, I discovered that David was *Samuel*. David scored a 93 percent. However, he was in pure rage when he received his score, and he confronted me after the class ended. He stated that he had the answer on the page, but I did not give him credit for it. Naturally, I asked to see his test.

When I saw his test, I was horrified for a second time. On his particular page, there were letters and numbers everywhere on the paper. I could not make sense of anything that David wrote. His paper looked as if 12-blindfolded graffiti artists wrote over every inch of the paper in different directions! And, at the very left top corner, I finally saw the answer: *"x = 3."*

It was obvious that David had chaos in his mind, and it was reflected on the paper. To solve the problem at-hand, I decided to strike a deal with him. I would give David points for that problem only under one condition: that he would watch the way I write each subsequent problem, and learn to write his solutions the same way. Surprisingly, David agreed.

We worked together for the future quizzes and tests. David and I sat down, and learned to write each math operation on separate lines. And, in the beginning stages of this process, he would write in English what he intended.

When David learned to write each math problem in a step-by-step way, I could understand his logic, and he could also identify when his own logic was incorrect. When you learn to write vertically, and each operation is represented on separate lines in a column, you can communicate to anyone your logic.

An example of this process is when you use the order of operations on:

(5(3)-40+10)/- 5+14/2

WRONG WAY TO SIMPLIFY:

In this example, the person has the original problem, but the work is done on the side and it is not clear how each number is obtained,

(5(3)-40+10)/- 5+14/2   15   40

                                                           -30

-5   7

Answer = 10

RIGHT WAY TO SIMPLIFY:

In this case, the problem is solved in sequential steps. These steps show the way each number is obtained, and each calculation is on a separate line. Also, the person solving this problem describes his/her actions on the right side of the problem in their own words. Notice that the equal signs are written directly above one another. This communicates a step-by-step process. This communicates to the reader, the logic that is in the mind of the math-solver.

$$(5(3)-40+10)/- 5+14/2$$

I am simplifying the numbers in the parenthesis now.

$$= (15-40+10)/- 5+14/2$$

$$= (-25+10)/- 5+14/2$$

$$= (-15)/- 5+14/2$$

$$= (3)+(7) \text{ } \textbf{\textit{I just divided to get this result.}}$$

$$= 10 \text{ } \textbf{\textit{I am adding the numbers now}}$$

In summary, the chart below shows the differences between *Samuel* and Ivan.

Samuel vs. Ivan

Samuel's Actions	*Ivan's Actions*
• Becomes ashamed because he is viewed as slow.	• Finds out how his brain works through the testing center and uses the methods taught to learn math effectively.

• Sees letters inverted.	• Clears his mind from stress through exercise, prayer, or mediation and sees an eye doctor and/or testing center to verify the ways his eyes/mind operates.
• Has Attention Deficit Disorder & has a wandering mind.	• Takes the time to nourish his mind with the right food based on his Blood type and changes 80% of his diet to eliminate toxins that slow his brain.
• Cannot read the board, listen to the lecture and write the notes simultaneously.	• Asks the teacher to record the audio from the lecture, and write his notes. Or asks another student to take notes while he listens (without writing).
• Never understands the questions.	• Asks the teacher to identify how questions are worded, so he can understand what is expected in each math problem.
• He will write all over the paper without sequential steps. Or, writes sentences on top of one another.	• Writes out each step and writes vertically (up to down). Explanations are next to each step.
• Thinks the math is moving too fast for him.	• Sits with the teacher and within study groups to find ways to maximize his learning time.

TURN CRYSTAL THE CRITICIZER into ELLEN THE EXPLORER

CHAPTER 9

TURN CRYSTAL THE CRITICIZER INTO ELLEN THE EXPLORER

Within one week, I had two experiences with two people who were *Crystal* the Criticizer. These two encounters showed me how and why *Crystal* the Criticizer is always in a nonprogressive, never-ending cycle. The first story was quite funny, while the second was seriously annoying.

Consider my first encounter with a *Crystal*-type character. A science television network contacted me, inquiring about shooting a demo video with me as the potential science TV host. The entire video had to be completed within seven days. And, I was tasked with hiring a film crew, as well as securing the location, per the company's budget. The casting executives also asked me to find a male or female scientist to interview. The trick was to interview the scientist with pertinent questions, along with incorporating details about myself on film. After finding a film crew to work within my budget, I gained authorization to film on campus. Fortunately, I had a few scientists at my disposal. One of them happened to be a former classmate from my university. Since graduating from our Math & Science program, he had started his own computer training company. He had created a fun, computer game geared toward helping banking institutions staff companies wisely by using mathematics. I thought all would be terrific, until the day of the shoot.

During the shoot, my camera crew and I witnessed a Dr. Jekyll/Mr. Hyde experience with this scientist "transforming" into a control freak! He started telling the videographer how to film the interview. Then, he attempted to dictate to me how to ask him the questions. In addition, he went so far as to try changing the location of the shoot. He was a pure nightmare! Studying with him years prior, I thought highly of him because he seemed well-versed on numerous issues. However, that was the past and I had since gained my own knowledge and experience. During the course of the interview, my perception of him changed.

I realized that he had always been the same controlling, bossy person. But, I had transformed. Now, I was more aware, competent and confident in my own expertise. I knew my plan, and I recognized how to gain help from other experts to create a superb project. At that moment, I identified the computer scientist as a *Crystal* the Criticizer. Needless to say, when the film started rolling, I let him meet the *real* boss.

The second story happened within the same week. A young lady from my church discovered that I was attending a private event through my RSVP on a social networking site. Excited to attend, she invited herself and more than 10 people. When I saw her at the party, I was perplexed. Originally, I thought that I invited her. Then, I realized I didn't, and was further shocked and irritated, upon discovering she used my name to gain access into the party. Later, I discovered that she invited others. Naturally, I confronted her. I told her that her actions were inappropriate because she hadn't received permission to use my name. To my surprise, she didn't think anything was wrong with her actions! She claimed she could attend any gathering she desired simply because she and I are "friends." Of course, I became angered, and I indicated that her logic was faulty. Suddenly, I had another epiphany. She, too, was *Crystal* the Criticizer, and had no intent on changing her domineering ways. Nevertheless, she was no longer eligible for my friendship.

You may wonder what these two stories have in common. Both the scientist and my church friend exhibited fears, causing them to react in overbearing manners. If you consider their individual mindset, you can see where each felt as if he and she weren't *important enough*. The scientist did not think he was important enough to simply be present for the interview. Instead, he felt he needed to perform others' jobs. As for my church acquaintance, she thought she was not *important enough* if she was required to ask to attend the party. She also felt a dysfunctional need to invite herself to prove her self-worth. In each of these cases, the *Crystal-type* personalities failed to trust the people around them. Both examples also failed to become a beginner again, and enjoy this experience called *LIFE*. In short, with an arrogant attitude, we can let our fears take over, and in turn, we become *Crystal* the Criticizer.

In this chapter, we will learn to eliminate *Crystal's* controlling ways by understanding when she first lost trust. We will reprogram her dysfunctional thinking, and equip her with the tools to start over!

Crystal the Criticizer's control issues typically stem from a past experience, where he or she was let down by people who should have protected her. Or, she was forced into a caregiver-role too early in life, and consequently never had the opportunity to be a beginner. As a self-coping dysfunctional action, *Crystal* the Criticizer learns to micro-manage. This domineering personality even controls every aspect of a situation, where she has little knowledge and thinks that she knows best. But, in reality, she is grossly mistaken. Oftentimes, she doesn't know how to succeed in a new field. Sadly, she ends up failing miserably, because she has adopted a stubborn attitude that only attracts disappointment.

> *"Crystal the Criticizer does not trust her teacher, and does not know how to be vulnerable enough to step in a beginner's role."*

Crystal the Criticizer's actions in math classes are similar to the actions of the scientist and the church acquaintance. When *Crystal* the Criticizer faces a difficult educational experience, she may blame the people who are placed in a position to help. Frequently, when she is in a complicated position, she pretends to know how to handle the situation. In math classes, she says that the teacher doesn't teach well. Or, she claims that the information that she saw before is nothing like her previous education. Perhaps, she may think that she has a better way to do the math. *Crystal* may always feel that she has a better approach. With enough of these excuses over the years, people near her will begin to leave, because they are tired of being ridiculed. People begin to avoid *Crystal* the Criticizer, and as a result, *Crystal* the Criticizer has trouble trusting anyone. *Crystal* the Criticizer does not trust her teacher, and does not know how to be vulnerable enough to step into a beginner's role.

Every learning experience is unique, and we have to choose to completely place ourselves in the moment. Meaning, we have to fully accept that the learning process will sometimes be difficult. It is only when we are curious about the learning process itself that we are equipped with the tools and talent to succeed. If we can let go of our assumptions of how a situation should be, then we can open our minds to learn. In turn, the blame game ends.

Step 1: Know *Crystal's* Root Issue

Let us understand *Crystal* more, and know about the cases where *Crystal* tries to criticize others. When she feels that she may not be able to understand important concepts, *Crystal* dysfunctionally copes with the situation by either blaming others, or trying to control the experience. Any psychologist knows that blame is a way to discharge pain and discomfort.

If you recall, *Crystal* is a person who cannot handle being open and vulnerable in a learning situation. She may study on her own, as well as speak poorly about her teachers or tutors. She may also refuse to attend class. With her actions, she becomes rigid and overbearing. All of her potential joy from self-development is squashed. Consequently, her controlling habits are making her feel incompetent.

Secretly, *Crystal* desires to "let her hair down," and have child-like fun, as well as get a replacement teacher. She is angered by the thought of always being expected to know how and when to accomplish tasks. At the same time, she needs to feel that she is respected amongst her peers. So, she thinks that through her overbearing actions toward the teacher – control, criticize, and blame – such obnoxious behavior will earn self-worth and respect. *Crystal* truly fears that she has too much on her plate and will fail miserably with all of the responsibility that she thinks she holds. She believes that she can create her "dominance," but doesn't understand that the only control that she possesses is over her own thoughts and actions. *Crystal* needs help understanding why she needs to allow herself to become a beginner again.

When *Crystal* views math, she sees herself as unable to allow someone to teach her new concepts. Secretly, she is wishing to be carefree again and not have the burden of "omniscient." Secretly, *Crystal* has a fear that she will be expected to start completely over again if she truly immerses herself into something that is new. She attempts to control all actions in hope of feeling accomplished. There is good news: we can change *Crystal's* self-sabotaging, controlling actions. In order to do so, *Crystal* must ask herself some inspiring questions:

In order to know the root of *Crystal's* issues, let's ask these basic questions. If you are a *Crystal*, close your eyes (if possible), sit back, relax and voyage back in your memory:

Questions for *Crystal:*

At what age did you first lose trust in your teacher or parent? It can be within any event, math class, or unrelated scenario. What happened when you tried to get help from that person?

How did you feel at that very moment? Explain those feelings on paper or say it aloud.

Today, what part of that feeling reoccurs when you have a check-off list to accomplish?

When you thought about failing or not understanding math, explain how you felt when you started to blame others to avoid being perceived as incompetent.

What subsequent event(s) triggered your desire to learn on your own?

When you feel the teacher may not explain the material well, which one of the following do you do? Circle up to two:

- Hire Tutors

- Self-learn the information

- See Friends

- Stop Attending Class

- Complain to Friends

- Go to YouTube Teacher demonstrations

Are you a person with numerous responsibilities? Name the top five?

What has been your worst teacher experience?

Why did you choose to return/continue with school?

What frustrated you the most while you were in math classes?

Remember the very first event where your trust was destroyed. Recall that moment. Now, say aloud:

> "I cancel that moment in time. I forgive myself for choosing to accept fear in my past. I will no longer choose to be ruled by the fear that tries to creep into my mind. Fear, you are no longer welcomed into my life."

Now, decide to change the outcome. Return to that moment in time when your trust was betrayed. Pretend that you have nothing to fear. What would you say to the person that let you down? Say it!

Next, change the outcome. Pretend that you chose to start over with a fresh slate. Describe how different the outcome would have been.

Open your eyes, and come back to this moment. What did you realize?

Ellen is an explorer. She looks at life with a desire to learn something exciting and new.

Become Ellen the Explorer

Ellen is the opposite of *Crystal*. Ellen has a tendency to seek novelty and take risks, in order to learn. Ellen loves to seek out new experiences, go everywhere, see everything – she is never bored. Always up for an adventure, Ellen tends to seek excitement, discover new ideas, meet new people and put strange theories into practice. She is an extremely, independent person who understands the importance of experts' advice and guidance.

Ellen lives a rich and diverse life. Her natural curiosity attracts a variety of situations. She can create new ideas, practice innovative theories and generate fresh experiences by choosing to understand new skills in which she has never mastered. Ellen is always looking for enriching growth in her life. Her adventurous attitude makes her an excellent scientist – she has devotion for investigation and a job well done.

She is a very likeable person, who's bright, and possesses a happy attitude about life and the people around her. Ellen has a playful and open personality, and she is able to break away from routine. In class, she is an amazing thinker, and is known to be inquisitive. She has lived through numerous experiences. However, she remains focused on what is directly in front of her on the board.

Step 2: Divorce Fault & Re-Marry Fortune

Howard Carter was working in Egypt 15 years before discovering King Tutankhamun's tomb in the Valley of the Kings in 1922. He had no idea what he and his crew of archeologists would discover after eight years of digging. Small items such as a faience cup, and a piece of gold foil indicated that there was some type of fortune awaiting Carter and his crew. Fortune hunters program their minds to look for things that are new. Without stubbornness, they venture into the world ready to become rich with unique experiences that very few people encounter. In this case, Ellen explores. Ellen learns to turn *Crystal*'s thoughts into fortunes. Ellen corrects *Crystal*'s negative thinking by capitalizing on her special strengths. Per math thoughts, Ellen reprograms *Crystal*:

- *Crystal* thinks, "I don't feel like being in this class."

- Ellen Reprograms, "I will enjoy my time in class. And I will see something new."

- *Crystal* thinks, "The teacher doesn't know what he's doing."

- Ellen Reprograms, "I will ask the teacher many questions, so I can understand his approach."

- *Crystal* thinks, "I can do math better than the class explanation."

- Ellen Reprograms, "I wonder how my explanation relates to the teacher's. I will ask."

- *Crystal* thinks, "I can learn the subject better from a tutor, and not from the teacher."

- Ellen Reprograms, "My tutor will help me to understand the teacher's instruction style."

- *Crystal* thinks, "I want to do math my own way.

- Ellen Reprograms, "I will learn to do math in many different, correct ways."

- *Crystal* thinks, "I should get full credit on my tests because it's close enough to the correct answer."

- Ellen Reprograms, "I will ask about what I can improve upon to get a better score next time."

- *Crystal* thinks, "There are very few people who know what they are doing."

- Ellen Reprograms, "I will find those people who can show me how to succeed."

Ellen's Daily Affirmations:

It is okay to be a beginner again.

To feel vulnerable is to know that I am alive.

I can ask for help and still be respected.

My teacher is unique, and I will do my best to understand him or her.

I will choose to be thankful and happy despite any past fear.

I will enjoy my time in class.

I am excited about re-learning.

Step 3: Learn to Let Go & Master the Art of Beginning Over

In graduate school, I was *Crystal* the Criticizer. I had a big head. Because I was a rocket scientist, I thought that I was just as brilliant as the professors, teaching in my graduate school. I could gain new concepts easily, and I thought that I could understand concepts without spending much time on the teacher's explanations.

Then, during my first year of graduate school, I earned two C's in two of my classes. Now, in graduate school, a "C" is equivalent to a fail. And I was in danger of failing out of the graduate program. In fear, I began blaming my professors for giving me those grades. I started to have "a chip on my shoulder" because of this scenario. This negativity wasn't improving my grade point average. Luckily, I realized that I needed to develop another game plan.

I started to study with the two strongest mathematician groups in my program: the Russian and Chinese study groups. They had mastered learning, and through my interactions, I discovered several things. I learned how to effectively study in groups; I found out that my own fundamental math preparation was poor; and I learned that everyone in the class was expected to re-do the approach that the teacher used. I also learned that when I tried to solve the problems in my own way, I would receive poor marks.

My experience truly helped me to begin to move from a *Crystal* attitude toward an Ellen behavior. Slowly, I began to see that I was not able to master concepts alone. I saw that I needed to accept responsibility for my actions and ask for help from peers, tutors and most of all, my teachers.

Use the D.A.C.S. Approach

1. Decide to Trust.

To effectively be an Ellen personality, we must trust the people who are willing to show us how to succeed. When you look inside of yourself, you'll see you have the power to choose wisely. We can change our minds at any time. We are never locked into any situation, job, class or responsibility that we cannot ultimately change. The future is not set in stone, and we have a gift called free will. Once we wake up to this epiphany, we can begin to look at life with new eyes – with the eyes of a child who can see endless fortune ahead.

Typically, in the *Crystal* the Criticizer's pasts, we experienced challenging situations, where our trust was destroyed early in life. And now, we have difficulty trusting others. The feeling to control a situation seems simultaneously unnecessary and overwhelming. However, the way that we gain trust again is counterintuitive: we must learn to trust ourselves, "Will I take the risk to further this relationship, knowing that I cannot control the outcome?"

When we are in classes, we are choosing to be there. No, we are not forced.

Rather, we desire it. We are choosing to be present. We are choosing to allocate weekly time to master a new craft. By taking classes, we are essentially deciding to start a new relationship with another person, who has taken the professional responsibility to show us how to succeed. For *Crystal*, the start of any new relationship is frightening because the outcome is unknown. But, starting over gives us a clean slate, offering opportunities to trust that a bright outcome awaits.

2. Allow New Beginnings.

Do you remember when you first learned to tie your shoe laces? Or, when you learned to ride a bike? We had to learn how to do these actions. When we are children, everything is a new beginning and being a beginner is something that we never think about. We can be a beginner at learning a new school lesson, practicing a new sport or playing an instrument. As a child, we are excited about learning and we have no embarrassment. We are expected to be inexperienced.

As we age, something changes. Somehow, we get this notion that we are expected to know everything. Suddenly, being a beginner isn't desirable. *Crystal*, often feels embarrassed about her (his) ignorance. While beginning to develop expertise in other areas, the thought of being a beginner in another area makes *Crystal* uncomfortable. But, just like when we were kids, new beginnings mean new growth.

There is a three-step process to choosing to be a beginner again. It is similar to the following:

1) First, we feel like an idiot.

2) Next, we make a decision to start over.

3) Then, we decide to capture our growth as milestones.

> *"Don't be afraid to be a beginner again! "*

When we finish these three steps, we will develop new expertise and leadership skills. However, this process may make us uncomfortable at first. But, through the process of choosing to start over, opportunities begin to appear in our lives. And, we can start to rediscover joy. Journaling is also a good way to track our progress. Then, we can reclaim the joy we first experienced as children.

3. Clarify the Past.

It never fails. Every semester, there is a student in my class who, despite what I've taught them, insists on solving math problems the same way they learned it prior to my class. Nine times out of 10, they make mistakes on homework and fail the tests. This person made the biggest possible error. They failed to simply ask for clarification.

When you learn new things, you must be open to new methods of learning familiar concepts. Let me describe this theory further. Let's say that you need to deposit a check into your banking account. For years, you have visited the branch to deposit the check. But now, a friend tells you that you can deposit the check into an ATM machine. Is the outcome any different? No. It is the same result, just different techniques of obtaining the same outcome.

Math is exactly the same. Oftentimes, we must gain a particular solution. We have to solve for a variable, find the point of intersection, find the inverse, etc. There is always more than one way to solve a problem.

However, it is our responsibility to ask for clarification about the best approach, and to understand the difference between the previous processes verses the process now. You can find ways to ask the instructor about past math methods of learning, and then ask for more clarification.

> *"There is always more than one way to solve a problem."*

In order to find the best method to solve problems, we must first ask the teacher to explain the fundamental theories between each approach. To do this, follow these steps:

- Write out a particular math problem on a sheet of paper, and solve it the way that you originally learned it.

- Then, solve the same problem the way that the teacher solved it in class.

- Next, ask the teacher to explain why these two methods are the *same*.

- Request your teacher to review another problem to practice the new approach.

With these steps, you will gain clarification, and a better understanding of why these two methods offer the same answer. Remember, you are not expected to know the correct answer immediately. But, you are expected to know the appropriate methods to solve the problem.

4. See Different Points of Views.

Once you see a situation from another perspective, it becomes easier to accept the situation objectively. Frequently, we must discover the teacher's point of view, instead of operating solely based on our own knowledge. It's unwise to place our time and effort into a specific area without ever discovering the teacher's technique for solving problems. The teacher may believe that we are not trying to learn, or participate in the learning relationship. Because there is no one-on-one communication with the teacher, we end up trying to control the outcome and the results are never pleasant. However, when see the teacher's point of view, we allow the teacher to see our view as well. Learning becomes an interaction, and not a dictatorship.

In order to look at math from the other points of view, you must take the following three steps:

- **Write down your questions when they arise.**

When you immediately have a question or concern, you must write down the question and get it out of your mind. If you don't get the idea out of your mind, and onto the piece of paper (or electronic device), your mind can start to play tricks on you. And, you can become frustrated and angered. To combat this, find a little pocket-sized notebook or electronic device to write out questions when they come to you.

- **Select a time to discuss these questions with the person.**

You must have your questions answered. Answers are part of the learning process. Frequently, you may question why one approach is not used to solve problems. Until this question is answered, you cannot effectively learn. To verify your questions are answered, you need to select a time per week to discuss these problems with the teacher. You need to ask about the old method, compared to the new way. And, you must ask why the results will be the same or different. In doing so, you gain clarity, and you master problem-solving.

- **Place yourself in the other person's shoes to see the situation from their view.**

You can gain peace once you step out of your own shoes and walk in another person's path. Oftentimes, you can jump to judgment without understanding the issue from the other person's side. For example, a student came to my class 30 minutes late every day. I pulled him aside and I asked what was wrong. I found out that the student awakens at 4:30 a.m., daily, taking three buses to attend my class. And the earliest time he could arrive was 30 minutes late. After discovering his situation, I could understand his need. So, I asked the class if anyone was willing to carpool, and a person volunteered. As a result, the student now could awake at 7 a.m., and arrive to class on time. If I did not walk in the student's shoes, I would not have understood his situation. How often have you tried to understand someone else's constraints?

We can move from a *Crystal* attitude in to an Ellen attitude. We cannot master concepts alone. As *Crystal* decides to learn something new, she learns to trust like Ellen. When *Crystal* learns to become a beginner again, she discovers Ellen's joy. When *Crystal* clarifies her past experiences, she gains Ellen's proficiencies. And, when *Crystal* sees constraints from an outside perception, she gains Ellen's great outcomes.

Crystal's Actions	*Ellen's Actions*
• The more she controls, criticizes, and places blame - her peers and authority figures begin to dislike her.	• The more she can explore, learn, and trust - she gains respect from everyone.
• Thinks she can learn the subject better from a tutor verses her teacher.	• Gains help from the tutor to understand the teacher's instructional style.
• Wants to do math her way.	• Takes advice from math experts.
• Clings to her old ways of solving problems.	• Understands why she previously learned a particular approach verses a new method. Allows herself to become a beginner again. Explores her possibilities.

LEAD THROUGH CHANGE: TIPS FOR LEADERS, PARENTS AND EDUCATORS

CHAPTER 10

LEAD THROUGH CHANGE: TIPS FOR LEADERS, PARENTS AND EDUCATORS

An Internet company asked me to create 10 math videos for their website. However, I wasn't given much insight or visual direction. Thus, I asked the company representative to provide more details. I was then informed they wanted instructional videos designed with "dual capabilities" for teachers and parents to not only learn, but also teach math to their students. At that moment, I realized the enormous need that exists in our society. Both parents and teachers need to know how to teach students without continuing to spread mathaphobia®.

This chapter is dedicated to the parents, teachers and leaders wanting to help their students. Divided into two sections, you will find; Basic Tips for Parents; and The 11 Effective Math Teaching Strategies for Teachers. Both sections provide leadership techniques and an overall philosophy for educational policy makers.

Basic Tips for Parents

When I was in high school, I witnessed a life-changing event. I sat in my mother's junior college math class. I am sure it was against campus policy for me to sit in the room without being enrolled, but Dr. Lee was a kind man who wanted to help his students. And my mother was a Beginning Algebra student, who desired to attend college. Yet, she needed daycare for me. Consequently, Dr. Lee let her bring me to class so she could continue attending school. While sitting in class, I saw Dr. Lee solve difficult problems, making the gesture appear effortless. I told myself that I wanted to be just like him in the future. I, too, wanted to be a math professor!

During this time, I was barely 13. And, I established a relationship with Dr. Lee throughout my teen years, because my mother swore to take all of his classes. My mother "claimed" that she couldn't understand any other math professor. So, she waited two years to take the next math class strictly taught by him. At the same time, my mother yearned to help me with my own math homework, but she did not have the comprehension skills. Thus, she often asked Dr. Lee to show me how to perform various math operations.

> *"As you more than likely have discovered, mathaphobia® is not just a fear of math, it is a general fear of problem-solving. "*

Most parents experience similar scenarios. They want to help their children in math, but either forgot math concepts themselves or never learned it. Some parents are fortunate financially, and can hire tutors to work with their kids every few days. However, the average American parent does not possess that capability. Parents may remember how to do the math once they sit with the students, but they have busy lives themselves. As a result, there are three basic tips that I will offer parents in these situations:

- **Know your Mathaphobia® and end it:**

As you more than likely have discovered, mathaphobia® is not just a fear of math, it is a general fear of problem solving. Once we learn to end our own fear of situations, we transform our lives, as well as those around us. We have all been a *Quincy* the Quitter, *Donna* the Over-doer, *Samuel* the Struggler, and *Crystal* the Criticizer not only in math, but in other situations, too. Almost all of us have experienced mathaphobia® in our lifetime. When you can choose to identify your dysfunctional way of thinking, and reject it, the fear no longer rules your life.

As a parent, your job is to be a leader in the home. Leaders must first do what we want others to do. Then, through example, others will follow. In the case of mathematics, we must first choose to identify and eliminate our own mathaphobia®. When we have transitioned into a fearless state, it is much easier to lead a loved one to do the same. To end mathaphobia®, we can identify the mathaphobia® by taking the online test on www. Mathaphobia.com. Next, we change the way we think about problems by replacing the fear with the affirmations given in chapters five through nine. With the right action plan, we can transform the mathaphobia® to become David the Determined, Sarah the Strategist, Ivan the Innovator and Ellen the Explorer respectively. When we end math fear, our frontal brain lobes restart. In other words, when fear is turned off, we can solve problems.

You can remove the stigma from failing if you have your students partake in this process, as well. As a result, you will be able to know how you think verses the way your child thinks. In doing so, communication opens, and students are able to say statements like, *"That was a Donna move,"* without having shame attached to their actions!

- **Master Plan B:**

You can help your child become a math problem-solver by mastering Plan B. Create backup plans for your child, in the event that your child cannot obtain math help from the teacher. Create a Master Plan B: Ask your child to name the people in his/her class who understands the concept. There are always three or more classmates that your student can contact. At the start of the next class, ask your child to gather the names and contact numbers of these classmates. You can arrange a time where the classmate can meet in a campus library to help your child. Or, you can suggest that they can set a time to do a video conference on their Smart phone or computers. This will help your child feel connected again.

A Plan C also exists. Ask your child to identify the math tutors provided on campus. Plan D: Ask your student about the other math teachers on campus. If one person is not able to help, choose the next backup path. Take plan G, Plan H, etc... Plan I....until the mission is accomplished. This process will train your child to become resilient and not give up when she needs help.

- **Practice The Admit & Sit Approach:**

Nothing is more powerful than your presence. And time is the commodity that is priceless. With this said, you can use your commodity to earn desirable results. Help your child by using your time. If your child comes to you with a math problem, you may have forgotten how to solve the problem. In this case, you can take the *Admit & Sit Approach*. Admit that you have forgotten, and sit with them to relearn the math. If you require help, have a tutor come in to teach YOU and your child. You will show your child that you are investing your time into his/her success

Calm Down and Liven Up: Tips for Math Educators

If you are a parent or an educator, you have faced the dilemma of teaching a fearful student how to learn math. This enormous task is the most difficult for a teacher. Not only do you have to teach the information at a quick rate, but you also have to dispel the fear that is accompanied with the student's confusion. Educators across the nation face this issue on a daily basis. Whether home-schooling or teaching packed classrooms: one mission is necessary.

You must maximize the information in which a student acquires.

One day, I sat down to analyze what I do to successfully help others learn math quickly. To my surprise, I realized that I used psychological approaches to calm a person into rejecting fear and moving toward creative-thinking. I have been teaching math since 1993. Over the years, I had been successfully, albeit on a subconscious level, using certain approaches on a regular basis that have proven to be winning. Such methods have also helped engaged students' interests in learning mathematics again.

Trick #1: Create Calm, Patient Settings

Oftentimes, a person becomes agitated when screaming or yelling exists within an educational environment. The converse is also true. Frequently, when a person has slow movements, expresses himself calmly and with distinctive pausing, a person becomes tranquil. This technique is vital in math education. Math educators are most effective if they speak slowly, and use pauses between words as a tool to help provide the student time to digest information. When an instructor has a calm and collective approach, while teaching, a student builds trust in the teacher: and thus, in the math subject matter.

This patient setting allows the brain to end the fight-or-flight hormonal response, and opens the creative thought in the frontal brain lobes, which is conducive to math education.

o   Speak slowly and pause between words.

o   Use slow body movements.

Trick #2: Get Students to See the "Big Picture"

Recently, I took up horseback riding. Strangely enough, I was introduced to the sport at a polo match. The polo instructor suggested that I take polo lessons. I had never rode a horse before, but he said to take the class and he could teach me how to ride a horse. I took the class, and loved polo. However, I saw that I truly needed to learn how to ride a horse to be an effective polo player.

So, I enrolled in another class: a beginning horse riding group lesson. The new teacher showed me how to "post, trot, and to canter." Cantering is basically, "horse jogging." Within a lesson, I tried to canter without her oversight, and she quickly told me to stop. Before I could ask, "Why," She started sharing the "big picture" with me. The horseback riding instructor said that she did not want me to canter with other more advanced students in the arena. The instructor told me to picture myself as a "bicyclist on a freeway." When I understood this fact, I found joy in taking my time to learn cantering. She was a great teacher because, she explained the big picture, and gave a practical example that I could understand.

When teaching math, ask the students if they can see "the big picture." And, share everyday examples of how the particular process is used outside of math class. When the students see its real-life application, they learn it deeply.

o   Ask questions for understanding.

o   Use real examples of how the math is applied in everyday situations.

Trick #3: Translate Math into English

Les étudiants doivent savoir le vocabulaire de mathematiques. No your eyes are not playing tricks on you. You just read a sentence in another language. Your possible discomfort and slight uneasiness upon seeing unfamiliar words are the same emotions numerous students experience when trying to comprehend math. Countless need a translation system so they can understand the ideas being expressed. While teaching about math formulas, write down its meanings next to them. Also, use dictation techniques. Say a sentence in English, and ask the students to translate it into mathematics. Then, say a sentence in math and ask the students to write it in English. Students always seem to get a kick out of this exercise when you do it for fun.

Are you wondering what the French statement said? If so, the answer is: The students need to know math vocabulary.

- o Translate the math into English and English into mathematics.

- o Have the students try to write down what they are saying in math language, so they are familiar with writing equations.

Trick #4: Reinforce Correct Thinking

The quickest way to build shame with a student is to say, "You are wrong." This is like stabbing the student in the heart. No one ever likes to hear that they are wrong. However, everyone enjoys hearing what they are doing right. Thus, knowing the right things to say is critical in mathematics education. When a student is incorrect say, "Not quite," "Instead," "Or, take this correct approach." These simple words help the student to continue on their critical thinking process to eventually reach the correct solution. Also, as an educator, you earn more of their respect in return.

- o Confirm every correct step by saying, "Good. Yes." And, when they get a step incorrect, I say, "Not quite. Let's try it again." "Everything was right until this point, what else can this be?

- o Nod your head for confirmation that they are on the right track.

Trick #5: Get into the Mind of Each Student, and Mimic their Communication Style

It is hard to read minds, unless we learn to observe people. I have noticed that people usually indicate what is on their mind. How do they let others know? Unless people are expert poker players, individuals usually show their thoughts by their facial expressions, and by their actions.

Here's another example. When a student continues looking at her paper instead of the lecture, chances are, she is completely lost and using her notes as a crutch. For another example, if a person has trouble writing math symbols yet has art covering his notebook, use pictures to show math steps. The list can go on. Try these techniques to get into your students' minds:

o   *__Listen__ to the student to determine how he or she thinks.*

o   If a person is very systematical, provide them with a step-by step process for each type of solution.

o   If a person has a short attention span, relate the math to a subject that catches his fascination.

o   If a person is very artistic, draw a picture for geometrical relationships.

o   If a person talks a lot, use talking as a method for demonstration.

Trick #6: Confirm Opportunity

Oftentimes, students may feel as if they will fail a math class, because they did poorly in the previous lesson or section. This belief is the biggest lie that students seem to accept. As an educator, you must instill in your students that they are able to ask any question about the previous class or lesson that they did not understand. In turn, you must be willing to answer any question that they previously did not understand. When students feel as if their past will not hold them back, they are more apt to delve into the current lesson with greater confidence. Confidence coincides with the frontal brain activity. This exchange of ideas allows each to understand the new material.

o   Confirm that just because a student did not previously know math, doesn't mean he won't be able to learn it now!

Trick #7: Teach Students how to Execute the Solution

Students want structure in their class. This is the secret desire that all students possess. They require a foundation and a plan. The best way that students can see how to execute problems correctly is to watch you. All students subconsciously mimic their educators. The following will help you smoothly execute your daily class routine:

o   Lay out a lesson plan with an agenda at the start of class.

o   Teach per a specific time allotment.

o   Make math steps and processes clear. Explain the process in English.

o   Execute a solution in more than one way.

o   Write math in sequential steps. Number the steps.

o   Define math vocabulary with terms and their corresponding math symbols.

o   Relate topics to real-life applications.

o   Provide positive reinforcement when answers are given. Responses such as "Not quite," "Great," "The answer is correct," will help boost your students' confidence!

o   Verify the solution. Check your work. Ask students to double-check your work.

o   Construct ways to include class members to participate in the lectures. You can ask each to construct a similar problem, and solve it. Or, ask if another approach can be used on the same problem.

Trick #8: Support the Philosophy that Each Question is Unique and Important

The biggest complaint that I hear coming from a student is, "The teacher did not want to answer my question."

Too often, students feel intimidated by educators. When we consider classroom order, this aspect is a good thing. However, when we consider how some can learn effectively, this is the most detrimental aspect responsible for embedding math fear. When maintaining order to a class, while simultaneously encouraging students, you can take a few extra steps. Always say, "No question is a dumb question." And, "Each question is valid and unique." After students respond with questions, state, "That is a good question."

If someone has a basic math question, simply answer it, and move on. Also, when a student asks a question, use their question as an opportunity to see how the student processes information. When a person asks a question, chances are that five-to-six people have the exact same question, but they may be too shy to ask. Lastly, when a person asks a question, challenge yourself and see if you can explain it in a clearer way. Your effort will always be appreciated.

> ## *Always say, "No question is a dumb question."*

o   If anyone has a question, I never laugh. Address each question as an excellent question.

o   View each question as an opportunity to gain a better understanding of how the student thinks.

o   Say, each question is valid and unique.

o   View the interrogator as a representative of others in the class who are thinking the same thought.

o   The more questions that are asked, demonstrates the student's effort to learn.

o   Use questions as an opportunity to teach in a more clear way.

Trick #9: Inform the Class that Perfection is Impossible!

In Aerospace, we always work in groups. We work in groups for two reasons. First of all, we've found that innovation frequently comes from collective brain power. Secondly, not one of the scientists is perfect. We scientists make mistakes. As a result, other scientists are assigned to double-check our work, in order to verify that logic and math calculations are correct. If rocket scientists are not perfect, then math students aren't.

For some strange reason, students think that they have to be perfect in mathematics, and they eventually disappoint themselves when they aren't. NEWS FLASH: No scientist is perfect, no student is perfect – no one is perfect. The reason we are here on Earth is to learn. And students are in the class to learn. If they know it already, they would not have to take the class. Please instill this IMPERFECTION message in your student. Let the student know that he or she *is* expected to make mistakes, and they are expected to make errors. (Yes, you read this correctly). Tell them, throughout the class, the goal is to make less and less errors. Tell them that some people may seem to execute the problem perfectly, but their performance is an individual one. Make it a game. Also, share that the final is a time when they can show their cumulative knowledge. On the final, they can show how they have finally minimized their mistakes to gain the "A."

o   There will always be a more efficient way to handle a problem, and there will always be someone better at math.

o   I teach that each student has a different understanding of the math language and a different skill level.

o   Imperfection = human.

Trick #10: Know that Every Brain is Different

When I was a child, there was a classmate in third grade that was a better artist. One day, she created this beautiful ballerina. I couldn't believe how artistic my classmate was considering her age. The ballerina had amazing life-like features: she included bone and muscle structures, along with the tutu, leotard and toe shoes! I went home to tell my mother about it. I was shocked to discover my mother's message. She said something I've forever kept close to my heart.

My mother looked at me and said, "There will always be someone better than you in some area. And you will always be better than someone else. You will never be the best, nor the worst. You must just be."

I did not fully understand the depth of her message until I was in my 30's. I realized that we each have distinct DNA and individual brains. Therefore, we have a certain skill set that works best with our brain structure. With this said, there are a million different techniques to solve the same math problem. There is no such thing as the "one way" or the right way to do math. Because we see things differently, we can see another approach, a faster way, or a cleaner notation. As long as the math principles and theorems are followed to gain the correct answer, each process is unique and beautiful, just like the ballerina drawings.

o   Teach that there are a million different ways of looking at math problems: there is no such thing as "one way" to do math problems, because each brain is different. However, there is usually a most efficient way.

o   Let them know that as long as math principals are followed, there will always be a correct answer to gain.

Trick #11: Teach that Math is the Language of Patterns in the Universe

THE SECRET: *We mathematicians, do not look at numbers, we search for patterns.* When we can recognize the pattern, we use math to predict the future and to see the past...some say that we give fortunetellers "a run for their money."

Every great mathematician knows the trick to math. In a massive array of numbers, we look for patterns before we ever consider the numbers. It is through the witnessed patterns, that we derive famous theorems and approaches. In short, mathematicians are excellent pattern finders. When we can see the pattern, we can use our tools to find the events that are probable in the future.

"We mathematicians, do not look at numbers, we search for patterns."

In math classes, we can use real-life examples to show patterns. You can even have your students find math-related articles, and see how many discover the same article. This tactic builds pattern recognition. When you begin to teach that math is about pattern recognition, you instill that math is a universal language used globally by every person. For, we are all apart of the same universe.

o   Show that math exists in a universal form despite a person's national origin.

o   Math represents the universal law of energy transfer from one state to another. Past, present and future energy can be modeled using patterns, which are communicated by the language of mathematics.

Whether an educator or parent, we now have the tricks to help create the calm, mathaphobia®-free environments conducive to learning. It is our job to empower every student with confidence. We must allow mistakes, as well as be patient. Once we engage students through interactive questioning, we witness information being digested. Letting students know that every question is important inspires them to commit to a learning relationship. Through communicating that math is a precise language we demonstrate that math provides guidelines. By working in groups, we can support community learning.

Thanks to Dr. Lee and his special teaching methods, my mother was able to successfully complete her required math courses and obtain her Associates Degree in Culinary Arts and Restaurant & Hotel Management. Because of Dr. Lee's patience and calm methods, my mother felt comfortable in learning this "foreign language" called math. Thankfully, he granted my mother permission to have me in the class, where I not only learned mathematics, but also acquired psychological teaching tactics. Such techniques I later incorporated into my own teaching style to help ease the vulnerable feelings that students attempt to overcome when learning math. Had it not been for Dr. Lee, I wouldn't have been a mathematics major at the university, and perhaps you wouldn't be reading this book! We can all be like Dr. Lee.

GET UP.
RISE UP.
MOVE FORWARD. GO.

CHAPTER 11

GET UP. RISE UP. MOVE FORWARD. GO.

"The secret of getting ahead is getting started." — Sally Berger

I am an accomplished Aerospace and Science expert in the media, but I had to learn to reject fear to move toward this point in my life. As you can see, my life has been about moving past fear in order to succeed and accomplish rare feats. On the other hand, during my early years, fear – and its cousins called Doubt, Worry, Anxiety, and Depression – ruled my thoughts. They controlled my feelings. Consequently, they dominated my actions! Considering my upbringing in the impoverished, gang- and drug-infested streets of South Los Angeles, Calif., fear could have destroyed my future. Sure, I could've easily dropped out of school. Like so many others, I could've remained hopeless: without dreams of escaping my unfortunate circumstances. I might've been the one who never made a significant difference across diverse communities and cultures. Yes, I could've fallen short of the calling that was placed on my life.

Thankfully, through several life-changing scenarios, I learned that I had a choice. I ascertained that I had an option: either continue holding onto fear, or do the unthinkable! Name the fear, discover its root, and ban the contract that I initially formed with the phobia. I chose to walk with courage – later learning to replace that gaping hole in my heart with true statements about my precious worth and destined future.

I had no idea that I was being undermined for years. Fear is a real, yet invisible being that we previously befriended when we didn't know any better. Fear is like a back-stabbing close friend, who gives poor advice. And its purpose is to undermine us at every progressive step. Our mission: to end the relationship once and for all.

My process to end my fears began when I entered college. While reading, a professor's handout indicating that 1-out-of 31,000 women of color would earn a Ph.D. in Math or the Sciences, I decided to be that one! In doing so, I had to recognize the fear that had previously ruled my life. As I prepared to earn my first degree, I saw others coping with the same math fear.

This phobia of mathaphobia® is real in countless people. I was afraid of math. I once failed Algebra and Geometry because I feared math. Even though I knew some of the basic math principles, I froze when asked to convey my understanding. During numerous high school classes, I was constantly scared that I would not understand the information presented. While tutoring in college, I encountered innumerable college students dealing with this same math fear. Sadly, they also failed classes. Our fears were the same four imposters described in previous chapters. And as I began dispelling the fear in others, I saw that there were four characters in total.

I finally decided to no longer accept my own math fear lies that I continuously believed. I meditated and asked for help. Then, my epiphany was shown. It is our job to recognize these imposters. I saw that we are the ones that invited fear into our lives. And, we established a contract with each fear-riddled character, allowing them to enter our lives, over and over again.

When you demand to see fear's face and know its name, you are empowered. You must see the fear for what it is worth – pure destruction. This allows you to no longer be fooled to think that it is your "friend." Next, you must demand to recall the traumatic time when you originally chose to accept the contractual lie which allowed each fear to return over and over again. When you can see the moment in time when you previously befriended it, you have the opportunity to sincerely ask for forgiveness for that moment. Open your mind for revelation from God or your Higher Source. Or, dig deep within and ask it from yourself. You then can cancel out that moment in time and start fresh again. This requires you to no longer invite that "so-called Fear friend" back into your life. Lastly, when you ask for the hardness in your heart to be removed and burned up, you now have the opportunity to request only goodness and prosperity from the Truth to enter your life forever.

> *"When you demand to see fear's face and know its name, you are empowered."*

I started to cast out fear from my own life. I saw my overdoing was from accepting a lie. I was compensating in order to deal with the broken family structure I had since my youth. I discovered that I was a struggler because I accepted a lie when my face became disfigured when I was 10. Controlling became an issue out of a lie I accepted when my mother had difficulty caring for us. At all of these stressful moments, I did not know any better. So I accepted a fear in order to cope. However, like you, I am now armed with the truth about fear. We are adults now. We are empowered to end any bad relationship.

As critical thinkers, we can override the fear within our brains. We can no longer solely be ruled by emotions. We can demonstrate that self-love is not what we feel, but it is how we act. We have the opportunity to start fresh and move forward to a new beginning, new future, and a new life. This is all true, if and only if, we decide to move forward.

In college, my main goal was to cast out math fear from myself and out of others. After repeatedly tutoring people, I began to recognize that there was a true fear of Math which I named, MATHAPHOBIA®. I began to observe the patterns of this fear. It was only then was I able to see the "big picture" of this virus. And, the fear was clever in its disguise. Thus, I began my quest to determine how to eliminate mathaphobia®, so I could help people live with freedom to be themselves again.

We have seen that in math, we have four main fears as observed in the characters; *Quincy* the Quitter, *Donna* the Over Doer, *Samuel* the Struggler, and *Crystal* the Criticizer. And as you have finally discovered, each of these fears are not only related to math, they are the dysfunctional coping techniques we used to deal with problems in our lives. If we learn which dysfunctional personality was pretending to be our friend, we can learn to be the true problem-solvers who make a difference in this world as exhibited in the characters; David the Determined, Sarah the Strategist, Ivan the Innovator and Ellen the Explorer.

When you take the *Mathaphobia*® Self-Test, you can place a face to your fear. Then you must completely know where your fear originated. If you are identifying with *Quincy*, you must determine what inspires you to quit and think that you are not worth the effort. If you are identifying with Donna, you must understand who you are truly trying to impress while you miss the mark. If you're identifying with *Samuel*, you must determine why you feel inferior to other students and why you think teachers do not want to help you. If you are identifying with *Crystal*, you must determine why you always feel as if you must know the answer and control how others view you.

> *"When you take the Mathaphobia® Self-Test, you can place a face to your fear."*

After this critical step, you must decide to forgive yourself, cancel that moment in time, and decide to no longer be that person again. And every day, tell yourself who you are and what you will do. If you choose to be David, you must be determined to not let anything or situations stop you from getting help. If you choose to be Sarah, you must decide to not impress anyone, and simply choose to see the path that will lead you to the best outcome without over-taxation. If you choose to be Ivan, you must decide to embrace the way that you see the world, and be courageous to share what you see. If you choose to be Ellen, you must choose to never place blame, but rather relinquish control to have fun as you learn from square one.

Next, when we decide to change, we must change the way that our brain operates. We must override the years of lies that we programmed our brains to accept. We must choose to take strategic actions to create new neuropath ways leading to the frontal brain lobes. Through repeated actions that are identified in this book, we can restructure our brain to move into a new, smart-er state: the state that solves problems. When we carry out the actions of the problems-solvers, David, Sarah, Ivan and Ellen, we can become people who can solve situations. We become people who are strong and able to face ad-versity. We become the people who we were made to be, and we conquer the fear by choosing new actions, and a new way of being.

After choosing to conquer the fear, and transforming into one of the four fearless people, we now must truly capitalize on the gifts in which we are given. We are each endowed with a distinctive mind. Per the Myers-Briggs personality typologies, we are empowered to know the way in which our mind best communicates and views situations. With this information, we are empowered with tools to study mathematics and communicate with people who think differently. This comprehensive psychological data also ensures us that our logic is valued as important.

Lastly, if you are an educator or parent, you have become aware of the tricks to help create the calm, mathaphobia®-free environments conducive to learning. This environment is created by the leader who takes ownership over empowering each student with confidence. Through communicating that math is a precise language, valuing incorrect answers and by making learning a collective group process, educators and parents can ensure that our society will have more math literate people in the future.

As you have read, it doesn't take a genius to be a rocket scientist. It took me to realize that the dysfunctional background that I faced as a child – severe poverty, broken household, gang violence, failure in math and environmental depression – had nothing to do with the future in which I could build for myself. The only difficulty I truly had to face was choosing to let go of the fears that I accepted from others. By helping others in the same situation, I witnessed a common theme of fear that halted our confidence. When I found out about this self-empowering mathaphobia® removal process, I learned to be brilliant in math. Ironically, this process has almost nothing to with math itself.

> *Being raised in a gang- and drug-infested neighborhood, where a fifth-grader like me gets stabbed in class is highly dysfunctional.*

Like I mentioned earlier, I would've never guessed that I would become a rocket scientist! My background was the farthest from what many consider impressive. Let's face it. Being raised in a gang- and drug-infested neighborhood, where a fifth-grader like me gets stabbed in class is highly dysfunctional! I can appreciate why many are often puzzled after discovering my accomplishments. Instead of becoming upset when asked, "How did you make it out of your environment," I now embrace a different response.

Now, I reply with this statement, "I learned to overcome mathaphobia®."

Through rejecting mathaphobia® and its self-sabotaging ways, I consequently built my talent to imagine myself as the person, who I always hoped
to be. My personal faith carried me through various trials and tribulations,

while grasping onto what I know is a universal principle: All things work
together in the end to create only good outcomes to those individuals who
sincerely seek "The Truth," and are called to transform this world! (See
Romans 8:28, KJV) Although I might not have been able to predict my future perfectly: however, I knew that a better future awaited me simply because I chose to work on releasing fear out of my life.

> *If I became a rocket scientist, given such early dysfunction, I am
> convinced that you can be anything you desire.*

If I became a rocket scientist, given such early dysfunction, I am convinced
that you can be anything you desire: if you choose to work on removing fear
from your life. In order to accomplish your dreams, simply take this information and apply it. The world is waiting to experience the gift that you will
bring to the universe – all mathaphobia®-free. Only good results await you in
your journey.

Now, get up. Rise up. Move forward. And blast off!

Bibliography

Arem, Cynthia A. *Conquering Math Anxiety. Book and CD*. Brooks/Cole Cengage Learning. Belmont, CA: 2010.

Brodie, Richard. *Virus of the Mind: The Science of the Meme*. Hay House Inc., New York, New York: 1996.

Contreras, Jose. Personal Interview 25 May 2010.

D'Adamo, Peter J. *Eat Right 4 Your Type*. G.P. Putnam's Sons. New York, NY: 1996.

Deming, W. Edwards. *The New Economics for Industry, Government, Education* – 2nd Edition. The MIT Press, 2000. (chapter 6, page 103).

Dyer, Dr. Wayne W. *Excuses Be Gone!: How to Change Lifelong, Self-Defeating Thinking Habits*. Hay House Inc., Carlsbad, CA and New York, NY: 2009.

Fox, Jennifer, M. Ed. *Your Child's Strengths: Discover Them. Develop Them. Use Them. A Guide For Parents and Teachers*. Viking, Penguin Group. New York, NY: 2008.

Mather, Kate. *"Space Shuttle Endeavour: Where to Spot the Shuttle"*. LA Times Blog. L.A. NOW. Sept. 20, 2012. 8:30 a.m. Los Angeles, CA. http://latimesblogs.latimes.com/lanow/2012/09/space-shuttle-endeavour-flyover-and-landing-viewing-tips.html).

Van Voris, Bob, *"Mark Madoff's Widow Blames His Suicide on Father Bernard Madoff in Book,"* Oct. 21, 2011. *http://www.bloomberg.com/news/2011-10-20/mark-madoff-s-widow-blames-former-father-in-law-for-husband-s-2010-suicide.html*. Bloomberg. Online. 1 July 12.

Waldron, Ben. *"Curiosity' Transmits Will.i.am Song from Mars." http://abcnews.go.com/blogs/headlines/2012/08/curiosity-transmits-will-i-am-song-from-mars/*. ABCNews.com. Online. Aug. 29, 2012.

http://en.m.wikipedia.org/wiki/Tom_Brokaw. Online. 8 July 12.

http://en.wikipedia.org/wiki/Los_Angeles_City_Hall. Online. 21 Sept. 12.

http://eu.guildwars.com/images/uploads/whiteMantle2.jpg. Online 15 July 11.

http://observer.com/2012/06/peter-madoff-pleads-guilty/. Online. 28 June 12.

http://www.aiatsis.gov.au/collections/exhibitions/ethnomathematics/docs/contents/m0005975_v_a.pdf. Online. 23 June 12.

http://www.boston.com/news/education/k_12/articles/2009/05/19/aspiring_teachers_fall_short_on_math/. Vaznis, James. "Aspiring Teachers Fall Short on Math; Nearly 75 percent fail revamped section of state licensing test." The Boston Globe. May 15, 2009. boston.com. Online. 23 June 10.

http://www.bloombergnews.com/news/2011-10-20/mark-madoff-s-widow-blames-former-in-law-for-husband-s-2010-suicide.html. Online. 15 Aug. 12.

http://www.californiasciencecenter.org. Online. Sept. 21, 2012.

http://www.cancunsteve.com/mayan.html. Online. 23 June 12.

http://www.eric.ed.gov/ERICWebPortal/search/detailmini.jsp?_nfpb=true&_ERICExtSearchValue_0=ED472097&

ERICExtSearch_SearchType_0=no&accno=ED472097. Online. 23 June 12.

http://en.m.wikipedia.org/wik/History_of_Catholic_education_in_the_United_States. Online. 18 Aug. 12

http://www.griffithobs.org. Online. Sept, 21. 2012.

http://www.hethert.org/musicians.jpg. Online. 23 Jun 12.

http://www.history.1900.s.about.com/od/1920s/a/kingtut.htm. Online. 6 Aug 12.

http://www-history.mcs.st-and.ac.uk/HistTopics/Mayan_mathematics.html. Online. 23 June. 12.

http://www.humanmetrics.com/cgi-win/jtypes2.asp. Online. 23 July 11.

http://www.mindfully.org/Food/Common-Food-Additives.html. Online. 23 July 11.

http://m.disneyland.disney.go.com/?CMP=KNC-D. Online. Sept. 21, 2012

http://www.mypersonality.info/personality-types/SJ-temperament/. Online. 23 July 11.

http://www.mypersonality.info/personality-types/NT-temperament/. Online. 23 July 11.

http:// www.mypersonality.info/personality-types/NF-temperament/. Online. 23 July 11.

http://www.mypersonality.info/personality-types/SP-temperament/. Online. 23 July 11.

http://www.poemhunter.com/poem/the-road-not-taken/. Online. 18 June 12.

http://www.navajocodetalkers.org. Online. 20 June 12.

http://suite101.com/article/indigenous-mathematic-systems-a2064. Online. 23 June 12.

http://www.medgadget.com/archives/img/Pajdic_Visionary_Anatomies.gif. Online. 15 July 11.

http://www.nytimes.com/2010/12/12/business/12madoff.html?pagewanted=all. Online. 1 July 2012.

http://en.wikipedia.org/wiki/List_of_space_shuttle_missions. Space Shuttle Launches. Online 1 July 12.

http://www.wptv.com/dpp/news/science_tech/mars-rover-curiosity-landing-landing-photos-videos-human-voice-makes-giant-leap-in-space-thanks-to-nasa. "Mars rover landing, Curiosity photos, video: NASA human voice makes giant leap in space." Mullen, Jethro, CNN. wptv.com. Aug. 31, 2012.

http://www.universalstudioshollywood.com/tickets/buy-a-day-get-a-year-free/. Online. Sept., 21, 2012.

Henriques, Diana, Baker, Al, Lattman, Peter, Robbins, Liz, Stellloh, Tim. *"A Madoff Son Hangs Himself on Father's Arrest Anniversary,"* The New York Times, Sunday, 12 Dec. 10, page A1.

Hutchinson, Lakshmi. *Personal Interview.* 21 June 2012.

Johnson, LouAnne. *Teaching Outside the Box: How to Grab Your Students by Their Brains.* Jossey-Bass Publishers. San Francisco, CA: 2011.

Kent Schocknick. *Personal Interview.* 15 July 2007.

King James Bible, 1611.

King, Ryane. *Personal Interview,* 14 July 2005.

Lemire, Jonathan and Fanelli, James. *"Mark Maddoff, son of Bernie Madoff, had it all before dad's billion-dollar Ponzi scheme was exposed," Saturday, Dec. 11, 2010. http:// articles.nydailynews.com/2010-12-11/local/27084086_1_ber. NYDailyNews.com.* Online. 1 July 2012.

Martha, Brooks. *Personal Interview.* 15 July 2007.

McKellar, Danica. *Kiss My Math: Showing Pre-Algebra Who's Boss.* Hudson Street Press, Penguin Group, New York, NY: 2008.

PatrickPretty. *"Peter Madoff Charged Criminally, Civilly; Bernard Madoff's Brother 'Enabled The Largest Fraud in Human History' and Gained Millions of Dollars, U.S. Attorney Preet Bharara Says," http:www.patrickpretty.com/2012/06/29/urgent-bulletin-moving. Peter-madoff-charged-criminally-civilly-bernard-madoffs-brother-enabled-the-largest-fraud—in-human-history-and-gained-millions-of-dollars-u-s-attorney-preet-bha/. Patrick-Pretty.com.* Online. 5:12 p.m. 29 Jun 12.

Platkin, Charles Stuart. *BREAKING THE PATTERN: THE 5 PRINCIPLES YOU NEED TO REMODEL YOUR LIFE.* Plume, Penguin Group, New York, NY: 2005.

*Ross, P. et al. *"National Excellence: A Case for Developing America's Talent."* Washington, D.C: Office of Educational Research and Improvement (ED), Programs for the Improvement of Practice. 1993.

Space Shuttle Launches, http://en.wikipedia.org/wiki/List_of_space_shuttle_missions.

About the Author

Olympia LePoint is best known for her role as an award-winning rocket scientist, a science entertainer, and university educator, striving to help people to overcome math fear. *Mathaphobia®: How You Can Overcome Your Math Fears and Become a Rocket Scientist,* is her debut self-help educational manual, designed to empower adults and students to ace math without doing more math problems! As an internationally-recognized leader within the Math and Sciences field, LePoint helped launched NASA's *Endeavour, Discovery, Columbia,* and *Atlantis* Space Shuttles. She won the 2004 Boeing Company Professional Excellence Award, and The 2003 National Black Engineer of the Year "Modern Day Technology Leader" Award.

LePoint has appeared in countless magazines and news publications, including recognition in 2010 as "The New Face for Math Literacy" for her

Mathaphobia® explanation on Oprah.com. She also appeared on *NBC News*, Dr. Drew's *Life Changers TV Show* in 2012, as well as *The Bret Lewis News Hour*, The Halogen Network, and on Christians OnDemand Episode(s). LePoint uses psychology in conjunction with math's creative-thinking to help individuals "Beat the Odds." This dynamic speaker is regularly featured as a guest Math and Science expert on TV and Radio.

She holds a Bachelors of Science Degree in Mathematics and a Masters Degree in Applied Mathematics from California State University at Northridge. As a CEO, keynote speaker and college educator, LePoint has channeled her passion into helping millions overcome fear and succeed in the Math and Science fields. She formed the national education program, *Olympia's End Mathaphobia*® Now, a system designed to end math illiteracy in America. Currently, LePoint serves as a Math and Science educational expert for TV, Radio, news publications, and other Social Media outlets.

Services

To acquire more information on Olympia LePoint as a keynote speaker, guest lecturer or to gain information on Mathaphobia® workshops and one-on-one coaching, visit:

www.OlympiaLePoint.com
www.Mathaphobia.com
www.facebook.com/OlympiaLePoint